Power Systems

For further volumes:
http://www.springer.com/series/4622

W. H. Tang · Q. H. Wu

Condition Monitoring and Assessment of Power Transformers Using Computational Intelligence

Springer

Dr. W. H. Tang
Department of Electrical Engineering
and Electronics
The University of Liverpool
Brownlow Hill
Liverpool L69 3GJ, UK
e-mail: whtang@liverpool.ac.uk

Prof. Dr. Q. H. Wu
Department of Electrical Engineering
and Electronics
The University of Liverpool
Brownlow Hill
Liverpool L69 3GJ, UK
e-mail: qhwu@liverpool.ac.uk

ISSN 1612-1287

e-ISSN 1860-4676

ISBN 978-1-4471-2626-3

ISBN 978-0-85729-052-6 (eBook)

DOI 10.1007/978-0-85729-052-6

Springer London Dordrecht Heidelberg New York

British Library Cataloguing in Publication Data
A catalogue record for this book is available from the British Library

Softcover reprint of the hardcover 1st edition 2011

Cover design: eStudio Calamar, S.L./Figueres.

Printed on acid-free paper

Springer is part of Springer Science+Business Media (www.springer.com)

To our Families

Preface

Power transformers are among the most expensive and critical units in a power system. The normal life expectance of a power transformer is around 40 years, and in many power systems the percentage of transformers operated more than 30 years is increasing due to the investment boom after the 1970s. As a result, the failure rate of transformers is expected to rise sharply in the coming years. Transformer failures are sometimes catastrophic and almost always include irreversible internal damage. Therefore, all key power transformers equipped in a power system should be monitored closely and continuously in order to ensure their maximum uptime. Generally, there are four main aspects of transformer condition monitoring and assessment, including thermal dynamics, dissolved gas, partial discharge and winding deformation, which should be monitored closely in order to determine power transformer conditions.

In recent years, rapid changes and developments have been witnessed in the field of transformer condition monitoring and assessment. Many research institutions and utility companies have their own condition monitoring and assessment guidelines for large power transformers. Most of such efforts are dedicated to developing accurate transformer models and reliable transformer fault diagnosis systems. These approaches are usually based upon empirical models, which are sometimes inaccurate and incomplete concerning abnormal operation scenarios. The major drawbacks are rooted in the inaccuracy of empirical thermal models, the lack of knowledge and evidence in dissolved gas analysis and intricate issues in winding deformation diagnosis. Nowadays, owing to the advance in computational hardware facilities and software data analysis techniques, the in-depth understanding of various phenomena affecting transformer operations has become feasible. With the use of advanced computational intelligence techniques, system operators are able to interpret correctly various fault phenomena and successfully detect incipient faults.

This book is dedicated to advanced model-based approaches to accurate transformer modelling and intelligent data mining techniques for reliable transformer fault diagnosis. It introduces three important up-to-date aspects of computational intelligence techniques to handle practical problems of transformer condition monitoring and assessment. These techniques include the evolutionary algorithms, the logical approaches and the cybernetic methods, which are employed for model parameter identification, fault feature extraction and classification and dealing with uncertainties for undertaking condition assessment of power transformers, respectively.

We wrote this book in belief that applying computational intelligence techniques to transformer condition monitoring and assessment would open the possibility of obtaining the maximum practicable operating efficiency and optimum life of power transformers, minimising risks of premature failures and generating optimal system maintenance strategies. This book is self-contained with adequate background introductions underlying analytical solutions of each topic and links to the publicly available toolboxes for the implementation of the introduced methodologies. It deals with practical transformer operation problems by analysing real-world measurements with a broad range of computational intelligence techniques. This book has presented many examples of using real-world measurements and realistic operating scenarios of power transformers, which fully illustrate the use of computational intelligence techniques to deal with a variety of transformer modelling and fault diagnosis problems. We hope that this book will be useful for those postgraduates, academics researchers and engineers working in the area of advanced condition monitoring and assessment of power transformers.

We would like to thank Dr. Almas Shintemirov for his contribution to chaps. 7 and 11, made during the period of his Ph.D. study undertaken at The University of Liverpool. We also wish to thank Dr. Kevin Spurgeon and Dr. Shan He for their contributions, made during the period of their Ph.D. studies undertaken at The University of Liverpool, to part of the achievements presented in this book. Special thanks are given to Mr. Zac Richardson and Mr. John Fitch of National Grid for supporting this work and providing useful discussions.

Special thanks also go to Anthony Doyle (the Senior Editor), Claire Protherough and Sorina Moosdorf for their professional and efficient editorial work on this book. Our thanks are also extended to all colleagues in the Intelligence Engineering and Automation Research Group, The University of Liverpool, for all assistance provided, and who have not been specially mentioned above.

The University of Liverpool, June 2010 Dr. W.H. Tang
Prof. Dr. Q.H. Wu

Contents

1 Introduction . . . 1
1.1 Background . . . 1
1.2 Main Aspects of Transformer Condition Monitoring and Assessment . . . 3
1.2.1 Thermal Modelling . . . 3
1.2.2 Dissolved Gas Analysis . . . 4
1.2.3 Frequency Response Analysis . . . 5
1.2.4 Partial Discharge Analysis . . . 6
1.3 Drawbacks of Conventional Techniques . . . 6
1.3.1 Inaccuracy of Empirical Thermal Models . . . 6
1.3.2 Uncertainty in Dissolved Gas Analysis . . . 7
1.3.3 Intricate Issues in Winding Deformation Diagnosis . . . 7
1.4 Modelling Transformer and Processing Uncertainty Using Computational Intelligence . . . 8
1.5 Contents of this Book . . . 9
1.6 Summary . . . 11
References . . . 12

2 Evolutionary Computation . . . 15
2.1 The Evolutionary Algorithms of Computational Intelligence . . . 15
2.1.1 Objectives of Optimisation . . . 15
2.1.2 Overview of Evolutionary Computation . . . 17
2.2 Genetic Algorithm . . . 18
2.2.1 Principles of Genetic Algorithms . . . 19
2.2.2 Main Procedures of a Simple Genetic Algorithm . . . 20
2.2.3 Implementation of a Simple Genetic Algorithm . . . 23
2.3 Genetic Programming . . . 24
2.3.1 Background of Genetic Programming . . . 24
2.3.2 Implementation Processes of Genetic Programming . . . 25
2.4 Particle Swarm Optimisation . . . 29
2.4.1 Standard Particle Swarm Optimisation . . . 30

2.4.2 Particle Swarm Optimisation with Passive Congregation 31
2.5 Summary 34
References 34

3 Methodologies Dealing With Uncertainty 37
3.1 The Logical Approach of Computational Intelligence 37
3.2 Evidential Reasoning 38
3.2.1 The Original Evidential Reasoning Algorithm 38
3.2.2 The Revised Evidential Reasoning Algorithm 45
3.3 Fuzzy Logic 48
3.3.1 Foundation of Fuzzy Logic 48
3.3.2 An Example of a Fuzzy Logic System 48
3.4 Bayesian Networks 50
3.4.1 The Bayes' Theorem 50
3.4.2 Bayesian Networks 51
3.4.3 Parameter Learning to Form a Bayesian Network 52
3.5 Summary 53
References 53

4 Thermoelectric Analogy Thermal Models of Power Transformers 55
4.1 Introduction 55
4.2 Conventional Thermal Models in IEC and IEEE Regulations 56
4.2.1 Steady-State Temperature Models 56
4.2.2 Transient-State Temperature Models 57
4.2.3 Hot-Spot Temperature Rise in Steady State 57
4.2.4 Hot-Spot Temperature Rise in Transient State 60
4.3 The Thermoelectric Analogy Theory 60
4.4 A Comprehensive Thermoelectric Analogy Thermal Model 61
4.4.1 Heat Transfer Schematics of Transformers 61
4.4.2 Derivation of a Comprehensive Heat Equivalent Circuit 63
4.5 Parameter Estimation of a Thermoelectric Analogy Model 66
4.5.1 Heat Generation Process 66
4.5.2 Heat Transfer Parameter 66
4.5.3 Operation Scheme of Winding Temperature Indicator 67
4.5.4 Time Constant Variation in a Heat Transfer Process 67
4.6 Identification of Thermal Model Parameters 68
4.7 A Simplified Thermoelectric Analogy Thermal Model 68
4.7.1 Derivation of a Simplified Heat Equivalent Circuit 68
4.7.2 Hot-Spot Temperature Calculation 70

4.8 Summary 70
References 71

5 **Thermal Model Parameter Identification and Verification Using Genetic Algorithm** 73
5.1 Introduction 73
5.2 Unit Conversion for Heat Equivalent Circuit Parameters 74
5.3 Fitness Function for Genetic Algorithm Optimisation 75
5.4 Parameter Identification and Verification for the Comprehensive Thermal Model 76
5.4.1 Estimation of Heat Transfer Parameters 76
5.4.2 Parameter Identification Using Genetic Algorithm . . . 77
5.4.3 Verification of Identified Thermal Parameters Against Factory Heat Run Tests 79
5.4.4 Comparison between Modelling Results of Artificial Neural Network and Genetic Algorithm . . . 82
5.5 Parameter Identification and Verification for the Simplified Thermal Model 85
5.5.1 Identification of Parameters Using Genetic Algorithm 85
5.5.2 Verification of Derived Parameters with Rapidly Changing Loads 87
5.5.3 Simulations of Step Responses Compared with Factory Heat Run 90
5.5.4 Hot-Spot Temperature Calculation 92
5.5.5 Error Analysis 93
5.6 Summary 93
References 94

6 **Transformer Condition Assessment Using Dissolved Gas Analysis** 95
6.1 Introduction 95
6.2 Fundamental of Dissolved Gas Analysis 96
6.2.1 Gas Evolution in a Transformer 96
6.2.2 Key Gas Method 98
6.2.3 Determination of Combustible Gassing Rate 99
6.2.4 Gas Ratio Methods 99
6.2.5 Fault Detectability Using Dissolved Gas Analysis . . . 100
6.3 Combined Criteria for Dissolved Gas Analysis 101
6.4 Intelligent Diagnostic Methods for Dissolve Gas Analysis 102
6.5 Summary 103
References 104

7 Fault Classification for Dissolved Gas Analysis Using Genetic Programming . . . 105
7.1 Introduction . . . 105
7.2 Bootstrap . . . 107
7.3 The Cybernetic Techniques of Computational Intelligence . . . 108
7.3.1 Artificial Neural Network . . . 108
7.3.2 Support Vector Machine . . . 109
7.3.3 *K*-Nearest Neighbour . . . 109
7.4 Results and Discussions . . . 110
7.4.1 Process DGA Data Using Bootstrap . . . 110
7.4.2 Feature Extraction with Genetic Programming . . . 112
7.4.3 Fault Classification Results and Comparisons . . . 116
7.5 Summary . . . 122
References . . . 123

8 Dealing with Uncertainty for Dissolved Gas Analysis . . . 125
8.1 Introduction . . . 125
8.2 Dissolved Gas Analysis Using Evidential Reasoning . . . 126
8.2.1 A Decision Tree Model under an Evidential Reasoning Framework . . . 127
8.2.2 An Evaluation Analysis Model based upon Evidential Reasoning . . . 129
8.2.3 Determination of Weights of Attributes and Factors . . . 131
8.2.4 Evaluation Examples under an Evidential Reasoning Framework . . . 133
8.3 A Hybrid Diagnostic Approach Combining Fuzzy Logic and Evidential Reasoning . . . 138
8.3.1 Solution to Crispy Decision Boundaries . . . 139
8.3.2 Implementation of the Hybrid Diagnostic Approach . . . 141
8.3.3 Tests and Results . . . 150
8.4 Probabilistic Inference Using Bayesian Networks . . . 152
8.4.1 Knowledge Transformation into a Bayesian Network . . . 153
8.4.2 Results and Discussions . . . 156
8.5 Summary . . . 161
References . . . 162

9 Winding Frequency Response Analysis for Power Transformers . . . 163
9.1 Introduction . . . 163
9.2 Transformer Transfer Function . . . 165
9.3 Frequency Response Analysis Methods . . . 166
9.3.1 Low Voltage Impulse . . . 166
9.3.2 Sweep Frequency Response Analysis . . . 167

9.4 Winding Models Used for Frequency Response Analysis 168
9.5 Transformer Winding Deformation Diagnosis 168
9.5.1 Comparison Techniques 170
9.5.2 Interpretation of Frequency Response Measurements 170
9.6 Summary 174
References 174

10 Winding Parameter Identification Using an Improved Particle Swarm Optimiser 177
10.1 Introduction 177
10.2 A Ladder Network Model for Frequency Response Analysis 178
10.3 Model-Based Approach to Parameter Identification and its Verification 179
10.3.1 Derivation of Winding Frequency Responses 179
10.3.2 Fitness Function Used by PSOPC 180
10.4 Simulations and Discussions 180
10.4.1 Test Simulations of Frequency Response Analysis . . . 181
10.4.2 Winding Parameter Identification 181
10.4.3 Results and Discussions 182
10.5 Summary 183
References 183

11 Evidence-Based Winding Condition Assessment 185
11.1 Knowledge Transformation with Revised Evidential Reasoning Algorithm 185
11.2 A Basic Evaluation Analysis Model 186
11.3 A General Evaluation Analysis Model 187
11.4 Results and Discussions 188
11.4.1 An Example Using the Basic Evaluation Analysis Model 188
11.4.2 Aggregation of Subjective Judgements Using the General Evaluation Analysis Model 192
11.5 Summary 193
References 194

Appendix: A Testing to BS171 for Oil-Immersed Power Transformers 195

Index 197

Acronym

AD	Axial displacement
AHP	Analytic hierarchy process
AI	Artificial intelligence
ANNs	Artificial neural networks
ANSI	American National Standards Institute
BNs	Bayesian networks
BPA	Basic probability assignment
BOT	Bottom-oil temperature
CF	Clamping failure
CI	Computational intelligence
CIGRE	International Council on Large Electric Systems
CPD	Conditional probability distribution
CPT	Conditional probability table
CTEATM	Comprehensive thermoelectric analogy thermal model
DD	DRM diagnosis
DGA	Dissolved gas analysis
DNA	Deoxyribonucleic acid
DRM	Dörnenburg's ratio method
EA	Evolutionary algorithm
EPS	Expert system
ER	Evidential reasoning
FDR	Fisher's discrimination ratio
FEM	Finite element method
FT	Fourier transform
FFT	Fast Fourier transform
FL	Fuzzy logic
FRA	Frequency response analysis
GAs	Genetic algorithms
GP	Genetic programming
HB	Hoop buckling
HE-D	High energy discharge

HST	Hot-spot temperature
HT-H	High temperature overheating
HV	High voltage
IEC	International Electrotechnical Commission
IEEE	Institute of Electrical and Electronics Engineers
J-E	Judgment-evaluation
KGD	KGM diagnosis
KGM	Key gas method
*K*NN	*K*-nearest neighbour
OD	Oil directed
OFAF	Oil forced and air forced
OLTCs	On-load tap changers
ONAN	Oil natural and air natural
LE-D	Low energy discharge
LT-H	Low temperature overheating
LV	Low voltage
MADM	Multiple-attribute decision making
MIT	Massachusetts Institute of Technology
MLP	Multilayer perceptron
NEMA	National Electrical Manufacturers Association
NG	National Grid, UK
NW	Normal winding
PC	"Construction-based" (phase) comparison
PCA	Principle component analysis
PDA	Partial discharge analysis
PDs	Partial discharges
PG	Poor grounding
ppm	Part per million
PSO	Particle swarm optimiser
PSOPC	Particle swarm optimiser with passive congregation
RBF	Radial basis function
RBFN	Radial basis function neural
RC	"Time-based" (reference) comparison
RD	RRM diagnosis
RM	Residual magnetisation
RMS	Root mean square
RRM	Rogers ratio method
RWS	Roulette wheel selection
SCT	Short circuited turn
SFRA	Sweep frequency response analysis
SGA	Simple genetic algorithm
SGT	Super grid transformer
SPSO	Standard particle swarm optimiser
STEATM	Simplified thermoelectric analogy thermal model
SVM	Support vector machine

SUC	"Type-based" (sister unit) comparison
TCG	Total combustible gas
TEA	Thermoelectric analogy
TM	Thermal modelling
TOT	Top-oil temperature
TV	Tertiary voltage
WT	Wavelet transform
WTI	Winding temperature indicator

Chapter 1
Introduction

1.1 Background

The electricity supply industry is usually divided into three functional sections, including generation, transmission and distribution. Power transformers, on-load tap changers, circuit breakers, current transformers, station batteries and switch gears are the main devices of a transmission and distribution infrastructure that act together to transfer power from power stations to homes and business customers. These devices are critical assets, and if they were to fail that could cause power outages, personal and environmental hazards and expensive rerouting or purchase of power from other power suppliers. Therefore, these critical assets should be monitored closely and continuously in order to assess their operating conditions and ensure their maximum uptime. Particularly, large oil-immersed power transformers are among the most expensive assets in power transmission and distribution networks. It can raise or lower the voltage or current in an AC circuit, isolate circuits from each other and increase or decrease the apparent value of a capacitor, an inductor or a resistor. Consequently, power transformers enable us to transmit electrical energy over great distance and to distribute it safely to factories and homes. A transformer can fail due to any combination of electrical, mechanical or thermal stresses. Such failures are sometimes catastrophic and almost always include irreversible internal damage. Part of failures may lead to high cost for replacement or repair and an unplanned outage of a power transformer is highly uneconomical. As a result, as major equipment in power systems, its correct functioning is vital to enable efficient and reliable operations of power systems.

There are various causes of transformer failures during operations, such as electrical disturbances, deterioration of insulation, lightning, inadequate maintenance, loose connections, moisture and overloading. Many of these failure effects, however, will increase in probability due to the passage of time with age. It is inevitable that some faults will occur, so it is very necessary to monitor closely on-line and off-line behaviours of transformers. Typical transformer maintenance

W. H. Tang and Q. H. Wu, *Condition Monitoring and Assessment of Power Transformers Using Computational Intelligence*, Power Systems,
DOI: 10.1007/978-0-85729-052-6_1,

programs include continuous assessment tests, frequent dissolved gas analysis and oil quality tests, thermographic scanning of transformers and electrical connections and on-line monitoring of questionable units. In the last four decades, three monitoring strategies have been developed for transformer fault detection and diagnosis:

1. A variety of relays have been developed to respond to a severe power failure requiring immediate removal of a faulty transformer from service, in which case, outages are inevitable. This is the so called reliability-centred monitoring, which cannot detect incipient faults.
2. Various off-line tests can be applied to detect possible incipient faults, and these tests are usually undertaken with respect to a regular time interval. However, such a time-based monitoring strategy is labour intensive and not cost effective, which is also ineffective in identifying problems that develop between scheduled inspections.
3. There is a trend in the power industry to move from time-based monitoring to condition-based monitoring, which employs advanced fault diagnosis techniques for detecting on-line and off-line incipient faults. A condition-based monitoring program can supply information about unit conditions in real-time, process these information and then determine when maintenance should be performed.

All the three strategies have been investigated for many years. The condition-based monitoring strategy is related to a wide range of on-line condition monitoring applications, which include the detection of partial discharges and insulation degradation, winding deformation diagnosis, monitoring of dissolved gas evolution, classification of hazards and assessing thermal conditions [1–3]. On the other hand, off-line tests can only be employed to identify faults after a transformer outage or after a scheduled time interval. It can be seen that, condition-based monitoring of power transformers can open the possibility of obtaining the maximum practicable operating efficiency and optimum life of power transformers, minimising risks of premature failures and providing the potential to optimal system maintenance strategies [1–4]. A collection of off-line routine tests is listed in Appendix A according to the British Standard BS171 [5] for reference purposes.

In recent years, rapid changes and developments have been witnessed in the field of transformer condition monitoring and assessment. The performance and reliability of transformers can be improved greatly by employing advanced on-line and off-line fault diagnosis systems. Many research institutions and utility companies have developed their own condition monitoring and assessment systems for key power transformers. For instance, in 1995, Massachusetts Institute of Technology (MIT) developed an adaptive intelligent monitoring system for large power transformers [4]. Four large transformers in the Boston Edison power network were under continuous surveillance by this system, which could summon attention to anomalous operations through paging devices. The MIT model-based monitoring system includes a thermal module and a gas analysis module, which has a facility for examining raw and processed data during transformer operations, and that facility is accessible through a modem connected to a computer. There are also commercialised products for such purposes, e.g. an on-line monitoring system MS2000 designed by ALSTOM [6], which gives a comprehensive survey of power transformers covering various aspects, including oil temperatures, gas-in-oil contents, tap changers, cooling units, etc. Based upon the field bus technology, it enables power engineers to increase the lifespan of a transformer and reduce fault possibilities. Both the two systems provide a basic structure for conducting unit surveillance and assessment, through which it can be conceived that a model- and network-based system is one of the most important focusses in the research field of condition monitoring and assessment of power transformers.

Nowadays, owing to the advance in computational hardware facilities and software data mining techniques, the in-depth analysis of various phenomena affecting transformer operations has become feasible. With the use of advanced modelling and data mining techniques, researchers are able to understand fault phenomena and utilise relevant data for accurately detecting transformer faults. Based upon the authors' extensive research in the field of condition monitoring and assessment of power transformers, this book is dedicated to advanced model-based approaches to transformer modelling and intelligent data mining techniques for assessing transformer conditions.

1.2 Main Aspects of Transformer Condition Monitoring and Assessment

There are four main aspects concerning transformer condition monitoring and assessment, i.e. thermal modelling (TM), dissolved gas analysis (DGA), winding frequency response analysis (FRA) and partial discharge analysis (PDA).

1.2.1 Thermal Modelling

The normal operation life of a transformer is partially related to the deterioration of its insulation through thermal ageing, which is determined mainly by its daily cyclic loadings. Transformer loading guides give guidance for selecting appropriate transformer ratings for given loading and cooling conditions and particularly for conditions with loading ratios above the nameplate rating of a transformer. For oil-immersed power transformers, the International Electrotechnical Commission (IEC) loading guide 60354 [7] can be used, while IEC60905 [8] considers dry type transformers. In the Institute of Electrical and Electronics Engineers (IEEE) loading guide [9], the same calculation methods as reported in IEC60354 are adopted, which are also similar to the loading guides reported by the American

National Standards Institute (ANSI) [10] and National Electrical Manufacturers Association (NEMA) [11].

The development of an accurate thermal model is always regarded as one of the most essential issues of transformer condition monitoring. The generally accepted methods, reported by IEC [7] and IEEE [9], can be used to predict the zones of hot-spot temperature in a transformer as the sum of the ambient temperature, the mixed top-oil temperature rise above ambient and the hot-spot rise above the mixed top-oil temperature. The two steady-state temperature rises of top-oil and bottom-oil above ambient can be estimated separately. Comparisons between the measured and calculated transformer temperatures, referring to the IEC power transformer thermal models, were discussed in [1, 4, 12]. There are also a few improved thermal models rooted on the traditional thermal solutions. For instance, a real-time mathematical thermal model was presented in [13], which consists of several differential equations and takes into detailed account the influence of weather on thermal behaviours of a transformer. Since the model presented in [13] is more detailed than the ANSI/NEMA method, the results were expected to predict accurate transformer temperatures by considering the incident solar radiation and correlations for cooling operations. However, the conventional calculation of internal transformer temperature is not only a complicated task but also with underestimated temperatures obtained based upon some assumptions of operation conditions. An accurate and meaningful thermal model is highly desired in practice in order to deal with transformer thermal ratings.

1.2.2 Dissolved Gas Analysis

Oil-immersed power transformers are filled with a fluid that serves various purposes. The fluid acts as a dielectric media, an insulator and a heat transfer agent. The most common type of fluid used in transformers is of a mineral oil origin. During normal operations, there is usually a slow degradation of the mineral oil to yield certain gases that are dissolved in the oil. However, when there is an electrical fault within a transformer, gases are generated at a much more rapid rate. DGA is probably the most widely employed preventative maintenance technique in use today to monitor on-line transformer operations, and a number of DGA interpretation guidelines have been developed by different organisations, e.g. IEC60559 [14], IEEE C57.104-1991 [15], CIGRE TF 15.01.01 [16] and GB7252-87 [17]. By applying a DGA interpretation technique on an oil sample, dissolved gases can be determined quantitatively. The concentration and the relation of individual gases allow a prediction of whether a fault has occurred and what type it is likely to be. For nearly forty years, DGA and its interpretation have been a useful and reliable tool for monitoring conditions of oil-filled transformers and other oil-filled electrical equipment.

However, based upon the conventional DGA interpretation methods, it is an arduous task to determine malfunction types and oil sampling intervals, due to various fault conditions and other interfering factors. Moreover, determining the relationships between gas levels and decline conditions is a perplexing task, because of complex gas combination patterns. Many attempts have been made to tackle DGA interpretation problems with a few recent developed computational intelligence (CI) techniques, among which artificial neural networks (ANNs) are the most widely used fault classifiers for DGA. In [18, 19], an ANN was utilised to detect faults based only upon previous diagnostic results. Expert systems (EPS) combined with other CI techniques have also been developed for DGA, e.g. fuzzy sets and evolutionary algorithms [20, 21]. These techniques can evaluate ongoing transformer conditions and also suggest proper maintenance actions.

1.2.3 Frequency Response Analysis

Nowadays, the sweep FRA (SFRA) technique has received worldwide attention for transformer winding condition assessment gradually replacing the low voltage impulse (LVI) technique. FRA [22] is a very sensitive technique for detecting winding movement faults caused by loss of clamping pressure or by short circuit forces. Variations in frequency responses may reveal a physical change inside a transformer, e.g. winding movement caused by loosened clamping structures and winding deformation due to shorted turns. In industrial practice, FRA is one of the most suitable winding diagnostic tools that can give an indication of displacement and deformation faults. It can be applied as a non-intrusive technique to avoid interruptive and expensive operations of opening a transformer tank and conducting oil de-gasification and dehydration, which can minimise the impact on system operations and loss of supply to customers and consequently save millions of pounds in timely maintenance. There are several international standards and recommendations for testing power transformers using SFRA, e.g. DL/T 911-2004 [23], CIGRE WGA2.26-2006 [24] and IEEE PC57.149 (draft) [25].

Most utility companies own databases containing historical FRA data for large power transformers. For example, in National Grid (NG, UK) large transformers are tested regularly using SFRA in a frequency range up to 10 MHz. By comparing a frequency response measured during maintenance with a fingerprint measurement obtained at an earlier stage, FRA is widely employed by utility companies as a comparative method in the low frequency range of several tens of KHz to 1 MHz. Differences may reveal internal damages of a transformer, therefore inspections can be scheduled for repairing. However such a comparative method cannot quantify the change caused by a fault and locate it. It is therefore necessary to develop an accurate FRA modelling and reliable fault diagnosis approach to interpreting the physical meaning underneath the variation of FRA data, which is with considerable industrial interest. A wide range of research activities have been undertaken to utilise and interpret FRA data for diagnosing winding faults, mainly including the development of accurate winding models [26–30] and the elaboration of FRA measurement systems [31, 32].

1.2.4 Partial Discharge Analysis

Electrical insulation plays an important role in any high voltage power apparatus, especially power transformers. Partial discharge (PD) occurs when a local electric field exceeds a threshold value, resulting in a partial breakdown of the surrounding medium as reported by IEC60270 [33]. Its cumulative effect leads to the degradation of insulation. PDs are initiated by the presence of defects during its manufacture, or the choice of higher stress dictated by design considerations. Measurements can be collected to detect these PDs and monitor the soundness of insulation during the service life of a power transformer. PDs manifest as sharp current pulses at transformer terminals, whose nature depends on the types of insulation, defects and measuring circuits and detectors used. The conventional electrical measurement of PDs is to detect PD current pulses with a testing circuit. However, given that the experimental data always consist of PD signals, sinusoidal waveforms and background noise, the extraction of useful information from PD signals is a very difficult issue.

The detection of PDs can be performed by a variety of techniques, most commonly electrical [4], acoustical [34], optical [35] and chemical techniques [36]. There are three types of PD analysis methods, i.e. the time-resolved partial discharge analysis [38], the intensity spectra based PD analysis [37] and the phase-resolved partial discharge analysis [38]. Because of the special characteristics of PDs, traditional digital signal processing methods are not suitable for analysing PD signals. Other useful time-frequency tools, e.g. Fourier transform (FT) and Wavelet transform (WT), can be employed to analyse PDs [39] for de-noising, characteristic extraction and data classifications. On-line partial discharge calibration and monitoring for power transformers have been introduced by using a pulse injection through taps of high voltage transformer bushings [40]. In summary, most of the techniques used for PDA are to denoise PD signals and extract useful PD pulses, which are in the area of advanced signal processing. As it is out of the scope of the CI techniques focussed in this book, the PD research is not presented in this book.

1.3 Drawbacks of Conventional Techniques

1.3.1 Inaccuracy of Empirical Thermal Models

The generally accepted temperature calculation methods, reported in the IEC and IEEE guides [7, 9], can be employed to predict hot-spot temperatures (HSTs), top-oil temperatures (TOTs) and bottom-oil temperatures (BOTs). The guides give mathematical models for determining the consequence of different loading ratios using a set of equations with empirical thermal parameters. However, the conventional calculation of internal transformer temperatures with exponential equations is not only a complicated task but also leads to a conservative estimate, obtained on the basis of some assumptions of operation conditions [1, 4]. Moreover, these empirical equations are mainly settled on thermal profiles of a specific transformer, and this detailed information is not likely available or always varies with time. Its ability to predict transformer temperatures under realistic loading conditions is somewhat limited (e.g. the traditional model cannot account for the variations of ambient temperatures and thermal dynamics when a transformer's cooler is on or off). Therefore, the development of a more meaningful and accurate thermal model for transformers is always regarded as a very important issue.

1.3.2 Uncertainty in Dissolved Gas Analysis

As known, not all the combinations of gas ratios presented in a fault can be mapped to a fault type as described in a diagnostic criterion. Different transformer DGA diagnosis techniques may give varied analysis results, and it is difficult for engineers to make a final decision when faced with so much diverse information. It is also known that some DGA methods, such as the Rogers ratio method, fail to clearly identify faults in transformers in borderline cases, while other DGA methods can identify these cases. Therefore, the integration of the available transformer DGA diagnoses to give a balanced overall condition assessment is very necessary. Additionally, transformer diagnosis interpretations are carried out by human experts applying their experience and standard techniques, and many attempts have been made to refine decision processes used to guide DGA reviewers for evaluating transformer conditions. Such attempts include EPSs [20] and the analysis of data using ANNs [41] or fuzzy logic [21, 42], which are limited in their representation of DGA interpretation as a classification or pattern recognition problem. Moreover, different transformer test methods, i.e. TM, DGA, FRA and PDA, have different advantages and limitations making it difficult to discard one and select another. Therefore a more intuitive idea is to combine all the results derived from major test methods and integrate these information to form an overall evaluation. As test results are sometimes imprecise and even incomplete, a suitable information integration method is required to process DGA data for dealing with such uncertainties.

1.3.3 Intricate Issues in Winding Deformation Diagnosis

Among various techniques applied to power transformer condition monitoring, FRA is the most suitable one for reliable assessment for detecting winding displacement and deformation. It is established upon the fact that frequency responses of a transformer winding in high frequencies depend on changes of its internal distances and profiles, which are concerned with its deviation or geometrical deformation. Thus, the calculation of internal parameters plays an important part in accurate simulations of transformer winding frequency behaviours. Modelling of a real winding in order to obtain frequency responses, being close to experimental ones, is an extremely intricate task since a detailed transformer model must consider each turn or section of a winding separately. However, in industry practice it is not always possible to conduct additional tests for precise measurements of transformer geometry or insulation parameter estimation.

Various model-based fault detection techniques have been applied to provide continuous and unambiguous indications of transformer conditions [4]. Generally speaking, a model-based system is constructed with well-understood mathematical descriptions, the parameters of which can be identified by optimisation. There are a variety of parameter optimisation techniques, e.g. genetic algorithms (GAs), which have been utilised in a wide range of applications in power systems. It is desirable to employ a powerful optimisation algorithm to identify winding model parameters in order to reduce the difference between simulations of a winding model and corresponding FRA measurements. Therefore, it is feasible to establish a model-based approach to transformer winding simulations employing a parameter optimisation technique using real FRA measurements. Furthermore, at the present time, the interpretation of measured frequency responses is usually conducted manually by trained experts. It includes a visual cross-comparison of measured FRA traces with reference ones taken from the same winding during previous tests and/or from the corresponding winding of a "sister" transformer, and/or from other phases of the same transformer. These comparison techniques aim to detect newly appeared suspicious deviations of the investigated trace compared with historical reference responses. The appearance of clear shifts in resonance frequencies or new resonant points may characterise faulty conditions of windings. Therefore, it is also necessary to study the effect of various winding faults on frequency responses in order to establish effective classification criteria for interpreting FRA results accurately.

1.4 Modelling Transformer and Processing Uncertainty Using CI

As stated in the preceding sections, it is important to monitor conditions of off-line and on-line transformers continuously. The main advantages of continuous monitoring are illustrated as follows: prevention of failures and downtime of a transformer, transformer life extension leading to the delay of investment for a new transformer and optimal transformer scheduling by means of condition-based maintenance instead of time-based maintenance. With the development of recent technologies in computational intelligence and signal processing, continuous condition-based monitoring has always been on the research spot devoted to power transformers, e.g. on-line temperature monitoring, on-site dissolved gas analysis and accurate winding deformation identification. For instance, an on-line DGA surveillance unit can be mounted on a transformer, which can make proactive decisions based upon sampled DGA data and determine when an observed transformer should be repaired or replaced. However, there still exist many challenging problems in this research area, such as how to process and understand a large quantity of on-line data, how to tackle uncertainty issues arising from fault diagnoses and how to develop accurate transformer models and further to identify their parameters?

CI is an offshoot of artificial intelligence, which combines elements of learning, adaptation, evolution and fuzzy logic to create programs that are, in some sense, intelligent. There are mainly three branches of CI, i.e. the logical approaches, the evolutionary algorithms and the cybernetic techniques, which are employed in this book for modelling transformers, dealing with uncertainties and furthermore determining optimal maintenance strategies. Moreover, the development of a condition assessment framework is highly desirable aiming at a balanced condition assessment system concerning power transformers, which is able to aggregate the diagnostic information from TM, DGA, PDA and FRA. This book deals exclusively with oil-immersed power transformers, as very little preventative maintenance can be carried out on a dry-type transformer with the exception of keeping it clean and dry. In summary, this book is devoted to the development of model-based approach and utilise CI techniques to provide practical solutions for achieving advanced condition monitoring and assessment of power transformers.

1.5 Contents of this Book

This book is organised as follows.

Chapter 2 introduces fundamentals of three evolutionary algorithms employed in this book for parameter identification and feature extraction. Firstly the basics of GAs are presented, including types of GAs, fitness functions and GA operators followed by a discussion on the main concept of genetic programming (GP) including genetic operators, terminals and functions of GP. Finally, the foundation of particle swarm optimiser (PSO) is presented along with an improved PSO algorithm, i.e. a particle swarm optimiser with passive congregation (PSOPC).

Chapter 3 focusses on three mathematical theories dealing with uncertainties, i.e. the evidential reasoning (ER) theory, the fuzzy logic (FL) theory and Bayesian networks (BNs). Firstly, an ER algorithm, which can be implemented to deal with imprecise and incomplete decision knowledge for a multiple-attribute decision-making (MADM) problem, is presented on the basis of the Dempster-Shafer theory. Both the original ER algorithm and the revised ER algorithm are discussed in this chapter. Then, the essential concepts of BNs are introduced, which are graphical representations of uncertain knowledge for probabilistic reasoning. Finally the FL theory is briefly explained using a simple example.

Chapter 4 begins by introducing conventional thermal models of power transformers, which involve calculations of steady-state and transient-state temperatures at different parts of a transformer. Then, this chapter presents an equivalent heat circuit based thermal model for oil-immersed power transformers and a methodology for model development and simplification. Two thermal models are developed built upon the thermoelectric analogy (TEA) theory, i.e. a comprehensive thermoelectric analogy thermal model (CTEATM) and a simplified thermoelectric analogy thermal model (STEATM), which are described in detail. The two thermal models are established to calculate real-time temperatures of main parts of an oil-immersed power transformer.

Chapter 5 employs a simple GA to identify thermal parameters of CTEATM and STEATM using on-site measurements. A comparison study between the ANN modelling and the GA modelling is made concerning CTEATM. The simulations of heat run tests and calculations of HSTs using STEATM are discussed in detail. In order to verify the two thermal models, a number of measurement sets with different operating scenarios are employed in model simulations and comparisons.

Chapter 6 gives a brief literature review of DGA, including the theory of faulty gas evolution and a variety of practical gas interpretation schemes, then several widely used DGA classifiers are briefly introduced for a reference purpose, e.g. ANN, FL and EPS.

Chapter 7 is concerned with DGA data preprocessing and fault classification. An intelligent fault classification approach to transformer DGA, dealing with highly versatile or noise corrupted data, is proposed. Bootstrap and GP are employed to improve interpretation accuracies for DGA. Firstly, bootstrap preprocessing is utilised to approximately equalise sample numbers for different fault classes to overcome the lack of fault samples. Then GP is applied to extract classification features from DGA gas data for each fault class. Subsequently, the features extracted using GP are fed as inputs to three popular classifiers for fault classification, including ANN, support vector machine (SVM) and *K*-nearest neighbour (*K*NN). Finally, the classification accuracies of three combined classifiers, i.e. GP-ANN, GP-SVM and GP-*K*NN, are compared with the ones derived solely from ANN, SVM and *K*NN respectively.

Chapter 8 deals with the uncertainties arising from DGA diagnosis from three different angles, i.e. evidence congregation, crispy decision boundary and probabilistic inference. In the first part of this chapter, the original ER algorithm is employed to combine evidence with uncertainties derived from a diverse range of diagnostic sources. The methodology of how to transfer a transformer condition assessment problem into an MADM solution under an ER framework is presented. Several possible solutions to the transformer condition assessment problem utilising the standard ER algorithm are then discussed. The second part of this chapter presents a combined ER and FL approach in order to demonstrate the practicality of using fuzzy membership functions for generating subjective beliefs. Ideas taken from the FL theory are applied to soften fault decision boundaries employed by conventional DGA methods. This has an effect of replacing traditional "Fault" or "No Fault" crisp reasoning diagnoses with a set of possible fault types and an associated probability of fault for each. The final part of this chapter investigates a probabilistic inference approach to reinforcing the capability of the IEEE and IEC DGA coding scheme when processing DGA data. A graphical model is derived using the BN theory to perform probabilistic inference for handling DGA fault classification problems. The cases, which are unidentifiable by the IEEE and IEC DGA coding scheme due to missing codes, are processed with the constructed BN in order to verify the proposed BN approach.

Chapter 9 firstly introduces the technical background of FRA. Then the advantages and disadvantages of various winding models are briefly discussed. Finally, the conventional FRA interpretation methods are summarised with a discussion on fault features related to different winding fault types using FRA.

Chapter 10 proposes a model-based approach to identifying distributed parameters of a lumped-element model of a power transformer winding. A simplified circuit of a lumped-element model is developed to calculate frequency responses of a transformer winding in a wide range of frequency. In order to seek optimal parameters of the simplified winding circuit, PSOPC is utilised to identify model parameters. Simulations and discussions are provided to explore potentials of the developed approach.

Chapter 11 begins by transferring an FRA assessment process into an MADM problem using the revised ER algorithm. Subsequently, two examples of transformer winding condition assessment problems are presented using two ER evaluation analysis models, where the potential of the ER approach to combining evidence and dealing with uncertainties is demonstrated. In the case when more experts are involved in an FRA assessment process, the developed ER framework can be used to aggregate experts' subjective judgement and produce an overall evaluation of the condition of a transformer winding in a formalised form.

Appendix A lists a number of routine tests according to the British Standard BS171, which are usually conducted for transformer off-line tests.

1.6 Summary

In this chapter, the background of transformer condition monitoring and assessment is introduced firstly. Three transformer monitoring strategies have been compared with each other, including reliability-centred monitoring, time-based monitoring and condition-based monitoring. The four main aspects of condition-based diagnosis techniques are presented, i.e. TM, DGA, FRA and PDA. Then the drawbacks of conventional transformer diagnosis techniques are discussed, followed by the main objectives of the book. Finally the book outline is provided to give a clear view of the entire contents.

References

1. Provanzana JH, Gattens PR (1992) Transformer condition monitoring realizing an integrated adaptive analysis system. CIGRE 1992, Rep. No. 12-105
2. Leibfried T, Knorr W, Viereck K (1998) On-line monitoring of power transformers—trends, new development and first experiences. CIGRE 1998, Rep. No. 12-211
3. Kemp IJ (1995) Partial discharge plant monitoring technology: present and future developments. IEE Proc Sci Meas Technol 142(1):4–10
4. James L, et al. (1995) Model-based monitoring of transformers. Massachusetts Institute of Technology, Laboratory for electromagnetic and electronic systems
5. Heathcote MJ (1998) The J&P transformer book, 12th edn. First published by Johnson & Phillips Ltd, Newnes imprint, UK
6. ALSTOMT&D Control Ltd. (2001) Monitoring system MS 2000, ALSTOM product manual
7. International Electrotechnical Commission (1991) IEC60354: Loading guide for oil immersed power transformers. International Electrotechnical Commission Standard, Geneva, Switzerland
8. International Electrotechnical Commission (1987) IEC60905: Loading guide for dry-type power transformers. International Electrotechnical Commission Standard, Geneva, Switzerland
9. Transformers Committee of the IEEE Power Engineering Society (1991) IEEE guide for loading mineral oil-immersed transformer. IEEE Std. C57.91-1995, The Institute of Electrical and Electronics Engineers, Inc., 345 East 47th Street, New York, NY 10017, USA
10. American National Standards Institute (1981) Guide for loading oil-immersed distribution and power transformers. Appendix C57.92-1981, USA
11. National Electrical Manufacturers Association (1978) Guide for loading oil-immersed power transformers with 65°C average winding rise. Standard publication No. TR 98-1978
12. Radaković Z, Kalić DJ (1997) Results of a novel algorithm for the calculation of the characteristic temperatures in power oil transformers. Electr Eng 80:205–214
13. Alegi GL, Black WZ (1990) Real-time thermal model for an oil-immersed, forced-air cooled transformer. IEEE Trans Power Deliv 5(2):991–999
14. International Electrotechnical Commission (1978) IEC60559: Interpretation of the analysis of gases in transformers and other oil-filled electrical equipment in service. International Electrotechnical Commission Standard, Geneva, Switzerland
15. The Institute of Electrical and Electronics Engineers (1994) Transformers Committee of the IEEE Power Engineering Society, IEEE guide for the interpretation of gases generated in oil immersed transformers, IEEE Std. C57.104-1991. The Institute of Electrical and Electronics Engineers, Inc., 345 East 47th Street, New York, NY 10017, USA
16. Mollmann A, Pahlavanpour B (1999) New guidelines for interpretation of dissolved gas analysis in oil-filled transformers. Electra, CIGRE France 186:30–51
17. Bureau of Standards for the P.R.China (1987) GB7252-87: Guide for the analysis and diagnosis of gases dissolved in transformer oil. National Technical Committee 44 on Transformer of Standardization Administration of China
18. Liu YL, Griffin PJ, Zhang Y, Ding X (1996) An artificial neural network approach to transformer fault diagnosis. IEEE Trans Power Deliv 11(4):1838–1841
19. Griffin PJ, Wang ZY, Liu YL (1998) A combined ANN and expert system tool for transformer fault diagnosis. IEEE Trans Power Deliv 13(4):1224-1229
20. Lin CE, Ling JM, Huang CL (1993) An expert system for transformer fault diagnosis and maintenance using dissolved gas analysis. IEEE Trans Power Deliv 8(1):231–238
21. Islam SM, Wu T, Ledwich G (2000) A novel fuzzy logic approach to transformer fault diagnosis. IEEE Trans Dielectr Electr Insul 7(2):177–186
22. Dick EP, Erven CC (1978) Transformer diagnostic testing by frequency response analysis. IEEE Trans Power Apparatus Syst, PAS-97 (6):2144–2150

23. Bureau of Standards for the P.R.China (2004) Frequency response analysis on winding deformation of power transformers, DL/T 911-2004, The electric power industry standard of P.R.China
24. CIGRE Working group A2.26 (2006) Mechanical condition assessment if transformer windings using frequency response analysis (FRA). ELECTRA, 228:30-34
25. The Institute of Electrical and Electronics Engineers (2007) IEEE PC57.149/D4 Draft trial-use guide for the application and interpretation of frequency response analysis for oil immersed transformers, 2007 (Draft)
26. Rahimpour E, et al. (2003) Transfer function method to diagnose axial displacement and radial deformation of transformer windings. IEEE Trans Power Deliv 18(2):493–505
27. Rahimpour J, et al. (2002) Transformer modeling for FRA techniques. In: Proceedings of the IEEE power engineering society transmission and distribution conference, vol 1, ASIA PACIFIC, pp 317–321
28. Rudenberg R (1968) Electrical shock waves in power systems: traveling waves in lumped and distributed circuit elements. Harvard University Press, Cambridge, Massachusetts
29. Shibuya Y, Fujita S, Hosokawa N (1997) Analysis of very fast transient overvoltage in transformer winding. IEE Proc Generation Transm Distrib 144(5):461–468
30. Bjerkan E, Høidalen H (2007) High frequency FEM-based power transformer modeling: investigation of internal stresses due to network-initiated overvoltages. Electr Power Syst Res 77:1483–1489
31. Wang M, Vandermaar A, Srivastava KD (2005) Improved detection of power transformer winding movement by extending the FRA high frequency range. IEEE Trans Power Deliv 20(3):1930–1938
32. Jayasinghe JASB, Wang ZD, Jarman PN, Darwin AW (2006) Winding movement in power transformers: a comparison of FRA measurement connection methods. IEEE Trans Dielectr Electr Insul 13(6):1342–1349
33. International Electrotechnical Commission (2000) IEC60270: High-voltage test techniques—partial discharge measurements, International Electrotechnical Commission Standard, Geneva, Switzerland
34. Ming L, Jonsson B, Bengtsson T, Leijon M (1995) Directivity of acoustic signals from partial discharges in oil. IEE Proc Sci Meas Technol 142(1):85–88
35. Lazarevich AK (2003) Partial discharge detection and localization in high voltage transformers using an optical acoustic sensor. The Virginia Polytechnic Institute and State University, Ph.D. thesis, Blacksburg, Virginia, USA
36. Gulski E, Kreuger FH, Krivda A (1993) Classification of partial discharges. IEEE Trans Electr Insul 28(6):917–940
37. Gulski E (1995) Discharge pattern recognition in high voltage equipment. IEE Proc Sci Meas Technol 142(1):51–61
38. Kreuger FH, Gulski E (1992) Computer-aided recognition of discharge sources. IEEE Trans Electr Insul 27(1):82–92
39. Tsai SS (2002) Power Transformer partial discharge (PD) acoustic signal detection using fiber sensors and wavelet analysis, modeling, and simulation. The Virginia Polytechnic Institute and State University, M.Sc. thesis, Blacksburg, Virginia, USA
40. Farag AS, et al. (1999) On-line partial discharge calibration and monitoring for power transformers. Electr Power Syst Res 50:47–54
41. Zaman MR (1998) Experimental testing of the artificial neural network based protection of power transformers. IEEE Trans Power Deliv 13(2):510–517
42. Tomsovic K, Tapper M, Ingvarsson T (1993) A fuzzy approach to integrating different transformer diagnostic methods. IEEE Trans Power Deliv 8(3):1638–1646

Chapter 2
Evolutionary Computation

Abstract The basic approach to optimisation is to formulate a fitness function, which evaluates the performance of the fitness function and improves this performance by choosing from available alternatives. Most classical optimisation methods produce a deterministic sequence of trial solutions using the gradient or higher-order statics of the fitness function. However, such methods may converge to local optimal solutions. The evolutionary computation approach is a population-based optimisation process rooted on the model of organic evolution, which can outperform the classical optimisation methods for many engineering problems. The existing approaches to evolutionary computation include genetic algorithms, evolution strategies, evolutionary programming, genetic programming and so on, which are considerably different in their practical instantiations. The emphasis of this chapter is put on the biological background and basic foundations of genetic algorithm and evolutionary programming. As the principles of particle swarm optimisation are similar to that of evolutionary algorithms, the standard particle swarm optimisation algorithm and an improved particle swarm optimisation algorithm are also presented in this chapter.

2.1 The Evolutionary Algorithms of Computational Intelligence

2.1.1 Objectives of Optimisation

Before investigating the mechanics and power of evolutionary algorithms, which belong to the evolutionary approach of CI, it is necessary to outline the objective of optimising a function or a process, as in this book evolutionary algorithms are

W. H. Tang and Q. H. Wu, *Condition Monitoring and Assessment of Power Transformers Using Computational Intelligence*, Power Systems,
DOI: 10.1007/978-0-85729-052-6_2,

applied to solve engineering optimisation problems. Mathematical optimisation is the formal title given to the branch of computational science that seeks to answer the question "what is the best?", for problems in which the quality of any answer can be expressed as a numerical value. Such problems arise in all areas of mathematics, the physical, chemical and biological sciences, engineering, architecture and economics, and the range of techniques available to solve them is very wide. In simple words, optimisation concerns the minimisation or maximisation of a function. The conventional view about optimisation is presented well by Beightler, Philips and Wilde [1]:

> Man's longing for perfection finds expression in the theory of optimisation. It studies how to describe and attain what is the Best, once one knows how to measure and alter what is Good or Bad ... An optimisation theory encompasses the quantitative study of optima and methods for finding them.

The objective of an optimisation problem can be formulated as follows: find a combination of parameters (independent variables) which optimise a given quantity, possibly subject to some restrictions on allowed parameter ranges. The quantity to be optimised (maximised or minimised) is termed the objective function; the parameters which may be changed in the quest for the optimum are called control or decision variables; and the restrictions on allowed parameter values are known as constraints.

Generally speaking, an optimisation technique is mostly used to find a set of parameters, $x = [x_1, x_2, \ldots, x_n]$, which can in some way be defined as optimal. In a simple case this might be the minimisation or maximisation of some system characteristics that are dependent on x. In a more advanced formulation an objective function $f(x)$, to be minimised or maximised, might be subject to constraints in the form of equality constraints, inequality constraints and/or parameter bounds. For instance, optimisation of an engineering problem is an improvement of a proposed design that results in the best properties for minimum cost. In more elaborate problems encountered in engineering, there is a property to be made best (optimised) such as the weight or cost of a structure. Then there are constraints, such as the load to be handled, and the strength of steel that is available. Thus, optimisation seeks to improve performance towards some optimal points. There is a clear distinction between the process of improvement and the destination or optimum itself. However, attainment of the optimum is much less important for complex systems. It would be nice to be perfect, meanwhile, we can only strive to improve [1].

Conventionally, the general search and optimisation techniques are classified into three categories: enumerative, deterministic and stochastic (random). Although an enumerative search is deterministic, a distinction is made here as it employs no heuristics [2]. Common examples of each type are shown in Fig. 2.1 [1, 2]:

1. Enumerative schemes are perhaps the simplest search strategy. Within a defined finite search space, each possible solution is evaluated. However, it is easily seen this technique is inefficient or even infeasible when search spaces become large.

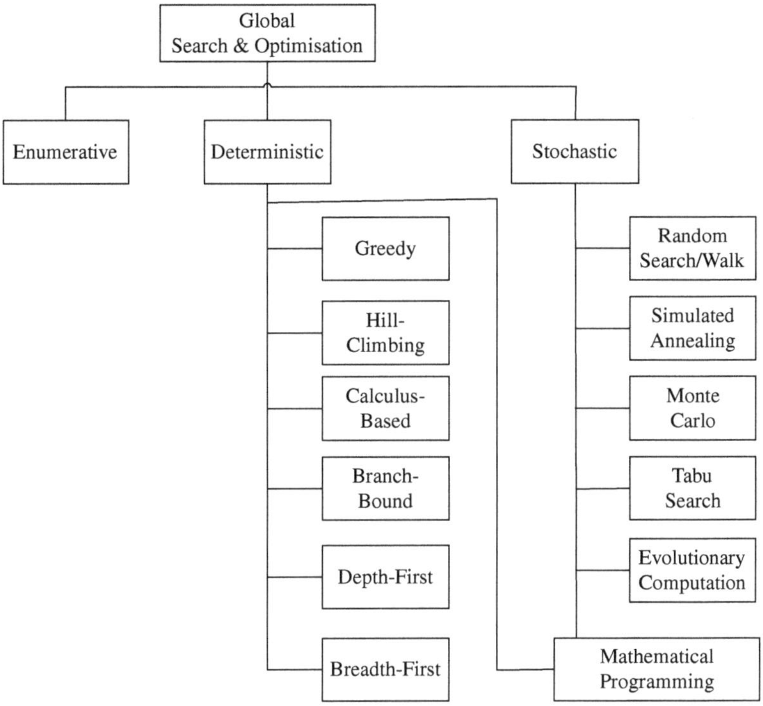

Fig. 2.1 Global search and optimisation techniques

2. Deterministic algorithms attempt to solve the inefficiency by incorporating problem domain knowledge. As many real-world problems are computationally intensive, some means of limiting the search space must be implemented to find acceptable solutions in acceptable time. Many of these are conceived as graph/tree search algorithms, e.g. the hill-climbing and branch-bound algorithms.
3. Random search algorithms have achieved increasing popularity, as researchers have recognised the shortcomings of calculus-based and enumerative schemes. A random search is the simplest stochastic search strategy, as it simply evaluates a given number of randomly selected solutions. A random walk is very similar amongst each other, except that the next solution evaluated is selected randomly using the last evaluated solution as a starting point.

2.1.2 Overview of Evolutionary Computation

Evolutionary computation techniques or evolutionary algorithms (EAs) work on a population of potential solutions in a search space. Through cooperation and competition amongst potential solutions, EAs can find optimal solutions more quickly when applied to complex optimisation problems. In the last forty years, the growth of interest in heuristic search methods for optimisation has been quite dramatic. The most commonly used population-based EAs are motivated from the evolution of nature. The subject now includes GA [1], GP [3], evolutionary programming [4], evolution strategies [5], and most recently the concept of evolvable hardware [6]. These algorithms stemmed from the very basic description of biological systems and were derived with a simple understanding of genetic evolution, which have shown the capabilities in solving optimisation problems of complex systems. EAs are classified as stochastic search algorithms for global optimisation problems, which have found many engineering and industrial applications [7, 8].

Different from these evolution-motivated evolutionary computation techniques, a recently emerged evolutionary computation technique, namely PSO [9, 10], is motivated from simulations of social behaviours. PSO shares many similarities with evolutionary computation techniques such as GAs, which is initialised with a population of random solutions and searches for optima by updating generations. However, unlike GA, PSO has no evolution operators such as crossover and mutation. In PSO, potential solutions, called particles, fly through a problem space by following the current optimum particles. In general, compared with GAs, the advantages of PSO are that PSO is easy to implement and there are few parameters to adjust. Recent studies of PSO indicate that although the standard PSO outperforms other EAs in early iterations, it does not improve the quality of solutions as the number of generations is increased. In [11], passive congregation, a concept from biology, was introduced to improve the search performance of the standard PSO. Simulation results show that this novel hybrid PSO outperforms the standard PSO on multi-model and high-dimensional optimisation problems.

All these related fields of research concerning GA, GP and PSO are often nowadays grouped under the heading of EAs, which is an offshoot of CI. EAs have been considered as general purpose parameter search techniques inspired by natural evolution models, which are appealing to many researchers in engineering. In this book, GA, GP and PSO are employed to identify model parameters and extract fault features for engineering problems. Detailed discussions on GA, GP and PSO are introduced in the following sections.

2.2 Genetic Algorithm

The application of GAs is one of the most important developments in the research field of EAs. GAs are excellent for quickly finding an approximate global maximum or minimum value, which explore a domain space with mutation and derive satisfactory results with selection and crossover. The two major problems using GAs are in converting a problem domain into genes (bit patterns) and creating an effective objective function.

2.2.1 Principles of Genetic Algorithms

GAs originated from the studies of cellular automata, conducted by John Holland and his colleagues at the University of Michigan. Holland's book, published in 1975 [12], is generally acknowledged as the beginning of the research of GAs.

Briefly, GAs require a natural parameter set of an optimisation problem to be coded as a finite-length string (analogous to chromosomes in biological systems) containing characters, features (analogous to genes), taken from some finite-length alphabets. For a binary GA, the binary alphabet that consists of only 0 and 1 is taken. Each feature is represented with different values (alleles) and may be located at different positions (loci). The total package of strings is called a structure or population (or, genotype in biological systems). A summary of the similarities between natural and artificial terminologies of GAs is given in Table 2.1.

A GA is generally recognised as a kind of optimisation method, which is different from the conventional optimisation techniques, e.g. gradients, Hessians and simulated annealing. GAs differ from the conventional optimisation algorithms in four aspects:

1. They work using an encoding scheme of control variables, rather than the variables themselves.
2. They search from one population of solutions to another, rather than from individual to individual.
3. They use only objective function information, not derivatives.
4. They employ probabilistic, not deterministic, rules, which do not require accurate initial estimates.

From the early 1980s the community of GA has experienced an abundance of applications, which spread across a wide range of disciplines. GAs have been applied to solve difficult problems with objective functions that do not possess nice properties such as continuity, differentiability, satisfaction of the Lipschitz Condition, etc. In recent years the furious development of GAs in sciences, engineering and business has lead to successful applications to optimisation problems, e.g. scheduling, data fitting and clustering and trend spotting. Particularly, GAs have been successfully applied to various areas in power systems such as power dispatch [13, 14, 15], reactive power planning [16, 17] and electrical machine design [18, 19].

Table 2.1 Comparison of natural and GA terminologies

Natural	GA
Chromosome	String
Gene	Feature
Allele	Feature value
Locus	String position
Genotype	Population
Phenotype	Alternative solution

2.2.2 *Main Procedures of a Simple Genetic Algorithm*

The GA used in this book is known as the simple genetic algorithm (SGA). The use of SGA requires the determination of five fundamental issues: chromosome representation, genetic operators making up the reproduction function, the creation of an initial population, termination criteria and an objective function. An SGA manipulates strings and reproduces successive populations using three basic genetic operators: selection, crossover and mutation [12]. The rest of this subsection describes each of these issues.

2.2.2.1 Solution Representation

In neo-Darwinism, we have a population of living organisms, i.e. the phenotype—coded by their deoxyribonucleic acid (DNA) and gene sequence—genotype. The genotype expresses its phenotype which competes in an environment. The competition drives the genotype to evolve a phenotype that performs best in an environment. The chromosomes found in living cells can be described as strings of many thousands of smaller units called alleles. There are only four different kinds of alleles. In the following example, we reduce the number of different kinds of alleles to 2 and the number of alleles in a chromosome to 10. Then, a simulated chromosome in an SGA scheme can be represented by a 10-digit binary number, e.g. 0010100111. The characteristic of an organism is determined by the particular sequence of alleles in its chromosomes. In this example, we parallel this concept by stating that the quality of any proposed binary number as a solution is determined by comparing it with an arbitrary ideal sequence which we are trying to find [20].

As mentioned previously, GAs are computer programs that employ the mechanics of natural selection and natural genetics to evolve solutions for solving an optimisation problem. In GAs there is a population of solutions encoded by a string. The representation of a possible solution as a string is essential to GAs as described in the above paragraph. A set of genes which corresponds to a chromosome in natural genetics is treated as a string in a GA. This algorithm, the most popular format of which is the binary GA, starts by setting an objective function based upon the physical model of a problem to calculate fitness values, and thereafter measures each binary coded string's strength with its fitness value. The stronger strings advance and mate with other stronger strings to produce offsprings. Finally, the best survives.

2.2.2.2 Selection Function

The selection of individuals to produce successive generations plays an extremely important role in GAs. This selection is based on the string fitness according to the "survival of the fittest" principle. A reproduction operator copies individual strings according to their fitness values. The strings with higher fitness values tend to have a higher probability of contributing one or more offsprings to the next generation. A selection method is required, which chooses individuals in relation to their fitness. It can be deterministic or stochastic, and the roulette wheel selection (RWS) method used in this study is discussed as the following [1, 20].

To reproduce, a simulated weighted roulette wheel is spun as many times as a population size. The selection for an individual, i, is stochastic and proportional to its fitness, f_i. It requires fitness values to be positive numbers, $f_i > 0, \forall i$, as each individual occupies a slice of a pie (hence a biased roulette wheel): f_i is the ith element of the total fitness $-\sum_{i=1}^{N} f_i$, where N is the population size. The probability of individual i to be selected is $P(i) = f_i / \sum_{i=1}^{N} f_i$. A uniformly distributed random number, R, is generated: $R \in U[0, 1]$. If R is between the cumulative probabilities of the ith and $(i + 1)$th individuals, then i is selected. This is repeated for the required number of replacements (usually N) for the next step.

2.2.2.3 Crossover Function

Nature modifies its code by crossing over sections of chromosomes and mutating genes, and GAs borrow this idea for this artificial algorithm. Once two parents have been selected for crossover, a crossover function combines them to create two new offsprings. The crossover operator operates in two steps following reproduction. First, each member in the newly reproduced string group is matched with another at random with a high probability p_c. Secondly, each pair of strings performs crossover with an exchange of each end part of strings at a certain position to generate a pair of new strings. An example of one point crossover is given below [20].

If a simulated weighted coin toss rejects crossover for a pair, then both solutions remain in a population unchanged. However, if it is approved, then two new solutions are created by exchanging all the bits following a randomly selected locus on the strings. For example, if crossover after position 5 is proposed between solutions 1100111010 and 1000110001, the resulting offsprings are 1100110001 and 1000111010, which replace their parents in the population.

2.2.2.4 Mutation Function

The mutation operator flips the code of certain digits of binary coded strings randomly with a small probability. For instance, if every solution in a population has 0 as the value of a particular bit, then a number of crossover operations may produce a solution with a 1 at a particular bit. This process could prevent strings from loss of useful genetic information, which usually results from frequent reproduction and crossover operations. In general, every bit of each solution is potentially susceptible to mutation. Each bit is subjected to a simulated weighted coin toss with a probability of mutation p_m, which is usually very low (of the order of 0.01 or less). If mutation is approved, the bit changes its value (in the case of binary coding from 1 to 0 or from 0 to 1).

2.2.2.5 Initialisation and Termination Criteria

GAs must be provided with an initial population as indicated previously. The most common method is to randomly generate solutions for the entire population. Since GAs can iteratively improve existing solutions, the initial population can be seeded with potentially good solutions, with the remainder of the population being randomly generated solutions. GAs move from generation to generation selecting and reproducing parents until a termination criterion is met. The most frequently used termination criterion is a specified maximum number of generations. Another termination criterion involves population convergence criteria. In general, GAs force much of an entire population to converge to a single solution. When the sum of deviations amongst individuals becomes smaller than a specified threshold, the algorithm can be terminated. The algorithm can also be terminated due to a lack of improvement in the best solution over a specified number of generations. Alternatively, a target value for an evaluation measure can be established based upon some arbitrarily acceptable thresholds. Moreover, several termination strategies can be employed in conjunction with each other.

2.2.2.6 Fitness Function

For engineering problems, GAs are usually employed to optimise model parameters, so that outputs of a model have a good agreement with reference values, subject to the minimal requirement that a function can map a population into a partially ordered set. A fitness evaluation function is independent of a GA, which depends on a particular problem to be optimised. In a simple term, the fitness function is the driving force behind a GA. A fitness function is called from a GA to determine the fitness of each solution string generated during a search. A fitness function is unique to the optimisation of the problem at hand; therefore, when a GA is used for a different problem, a fitness function must be formulated to determine the fitness of individuals. For many problems, a fitness value is normally determined by the absolute error produced by a GA individual with respect to a given reference value. The closer this error to zero, the better the individual.

Suppose o denotes the desired signal raw and the output raw of a GA individual is p. In general, the fitness can be calculated using an error fitness function or an objective function:

$$f = \frac{1}{n}\sum_{j=1}^{n} |p(j) - o(j)| \quad \text{or} \quad f = \frac{1}{n}\sum_{j=1}^{n} \sqrt{(p(j) - o(j))^2} \tag{2.1}$$

or a squared error fitness function:

$$f = \frac{1}{n}\sum_{j=1}^{n} (p(j) - o(j))^2, \tag{2.2}$$

where n is the number of output target samples.

2.2.3 Implementation of a Simple Genetic Algorithm

The SGA used in this book is implemented in binary coding, and its computation process is listed in Fig. 2.2:

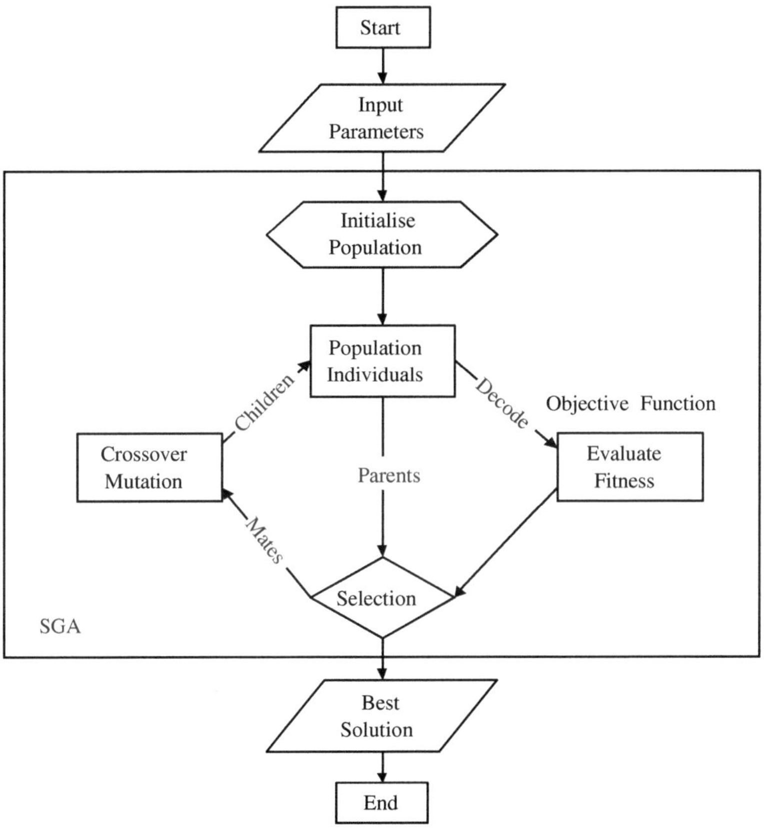

Fig. 2.2 A basic computation process of SGA

1. Initialise a population.
2. Evaluate the fitness of every string in the population.
3. Keep the best string in the population.
4. Make a selection from the population at random.
5. Crossover on selected strings with a probability p_c.
6. Mutation on selected strings with a probability p_m.
7. Evaluate the fitness of every string in a new population.
8. Make elitism.
9. Repeat (4) to (8) until a termination criterion is met.

The crossover probability, p_c, and the mutation probability, p_m, the size of the population and the maximum number of generations are usually selected as "a priori".

2.3 Genetic Programming

2.3.1 Background of Genetic Programming

GP, established by Koza in his fundamental book [3] upon the concepts of GAs, has become one of the most applied techniques in evolutionary computation.

The main difference between GP and GA is the representation of individuals in a population. Whilst GA encodes solution variants into fixed-length strings (chromosomes), GP has no such requirements, since a tree-structured (or hierarchical) representation of GP individuals holds an ability to evolve individual structures during a learning process, i.e. dynamically vary its size, shape and values. GP produces mathematical expressions as solutions. According to Langdon [21]:

> Genetic programming is a technique, which enables computers to solve problems without being explicitly programmed.

A complete GP process is typically a GA and repeats its operation sequence as listed in Sect. 2.2.3 [3, 22]. In order to run GP, several preliminary procedures are required to be undertaken [22]:

1. Determination of terminals and functions.
2. Definition of a fitness function.
3. Choosing GP parameters such as a population size, a maximum individual size, crossover and other probabilities, a selection method and termination criteria (e.g. maximum number of generations).

The population of GP individuals, being constructed as tree-structured expressions, is undergone by a procedure of fitness evaluation, which represents the individual survivability during a selection procedure. Then the fittest individuals, being chosen as parents for performing genetic operations, produce offsprings constituting new generations of a population. The process continues until a given termination criterion is met or simply a certain generation number is reached. The finally survived individual is treated as a variant of a desired solution. The tree structure of individuals allows GP to vary its size and shape, thereby, achieving a high efficiency in searching of a solution space with respect to what GAs are able to do [3, 22].

2.3.2 Implementation Processes of Genetic Programming

2.3.2.1 Terminals and Functions

GP generates an expression as a composition of functions from a function set and terminals from a terminal set. The choice of functions and terminals, which are collectively referred as nodes, plays an important role in GP since they are the building blocks of GP individuals.

Terminals correspond to the inputs of GP expressions, whether they are constants, variables or zero-argument functions that can be executed. Regarding tree-structured (or hierarchical) representations of individuals, terminals end a branch of a tree. In order to improve GP performance, an ephemeral random constant can also be included as a terminal [3].

Functions are chosen to be appropriate to a problem domain, which may be presented by arithmetic operations, standard mathematical, logical and domain-specific functions, programming functions and statements. In this book, only the mathematical functions listed in Table 2.2 are adopted for feature extraction using GP after numerous GP trials with different sets of functions are utilised.

2.3.2.2 Population Initialisation

The initial generation of a population of GP individuals for later evolution is the first step of a GP process. In general, the size of a newly initialised or reproduced

Table 2.2 A function set for feature extraction using GP

Symbolic name	No. of arguments	Description
Add, Sub	2	Addition, substraction
Mul, Div	2	Multiplication, division
Power	2	Involution
Sqr, Abs	1	Square, absolute value
SqrtAbs	1	Square root of absolute value
Exp, Ln	1	Exponent, natural logarithm
Sin, Cos	1	Sine, cosine
Arctan, Not	1	Arc tangent, invertor

GP individual is bounded by the maximum depth of a tree, i.e. by the maximum total number of nodes in the tree.

The depth of a node is the minimal number of nodes that must be traversed to reach from the root of the tree to the selected node and, correspondingly, the maximum depth is the largest depth being permitted between the root node and the outmost terminals of an individual [22].

In most cases, the initialisation of GP tree structures is implemented using the full or grow methods [3]. The grow method creates an irregular shape tree structure due to random selections of its nodes, whether it is a function or a terminal (except the root node being only a function). Thus, the maximum depth of a tree could not be reached until the terminal node is appeared, concluding the tree branch. As an example in Fig. 2.3a a tree-structured GP individual of a maximum depth of 4, calculating the following expression:

$$a(b - c) + \sin b, \tag{2.3}$$

is presented being initialised with the grow method. The terminals are variables a, b and c, whereas arithmetical functions +, $-$ and *sin* are the functions.

On the other hand, the full method generates tree structures by choosing only functions to build nodes in a tree branch until it reaches a maximum depth. Then only terminals are chosen. As a result, each branch of the tree is of the full maximum depth [22]. For instance, the tree in Fig 2.3b, representing the following expression:

$$ab - (b + c), \tag{2.4}$$

is initialised using the full method with a maximum depth of 3.

The ramped half-and-half method has also been devised in order to enhance the population diversity by combining both the full and grow methods [3]. Given the maximum depth d, a GP population is divided equally amongst individuals to be initialised having maximum depths 2, 3, ..., $d - 1$, d. For each depth group, half

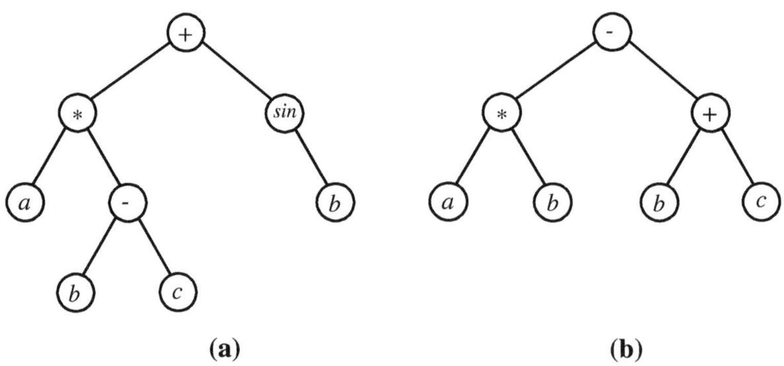

Fig. 2.3 Tree-structured GP expressions

of the individuals are generated using the grow method, and the other half by using the full method.

2.3.2.3 Genetic Operators

In GP the populations of hundreds and thousands of expressions are generally bred using the Darwinian principle of survival and reproduction of the fittest. Similar to GAs, three genetic operators, i.e. crossover, mutation and reproduction, are employed for this breeding, which are appropriate for generating a new offspring population of individual expressions from an initial population.

The crossover operation is used to create new offspring individuals from two parental ones selected by exchanging of the subtrees between parental structures as shown in Fig. 2.4. These offspring individuals are generally with different sizes and shapes to their parents [22, 23]. Mutation is operated on only one individual by replacing a subtree at a randomly selected node of an individual by a randomly generated subtree as shown in Fig. 2.5. The reproduction operation makes a direct copy of the best individual from the parental population and places it into the offspring population [22].

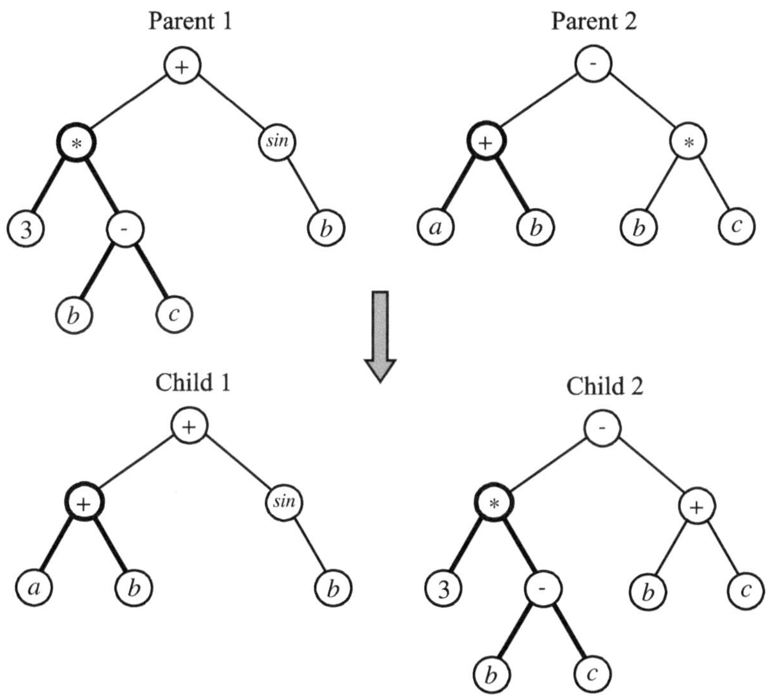

Fig. 2.4 Crossover of tree-structured GP expression

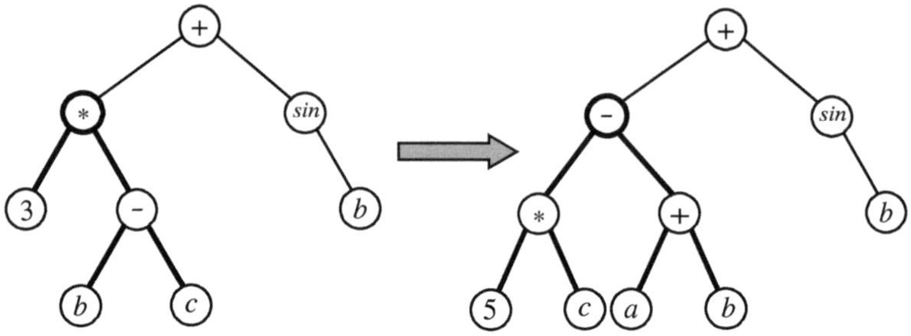

Fig. 2.5 Mutation of tree-structured GP expression

2.3.2.4 Fitness Function

Each individual expression in a population is evaluated in order to quantify how well it performs in a particular problem environment, which is represented by a fitness value. For a two-category classification problem, the Fisher's discriminant ratio (FDR) criterion is usually utilised as a fitness function. FDR is based on the maximisation of between-class scatter over the within-class scatter [24, 25]. Thus, each GP individual is evaluated according to its ability to separate particular classes of data by using the following equation [25]:

$$\text{fitness} = \frac{(\mu_1 - \mu_2)^2}{(\sigma_1^2 + \sigma_2^2)} - pN, \tag{2.5}$$

where μ_1, μ_2 and σ_1^2, σ_2^2 denote the mean values and variances of the two categories to be separated, respectively. p is a small value, e.g. 0.0005, which is introduced as a penalty to the fitness function depending on the number of nodes N of each evaluated individual. This allows a GP program to control the increase in the size of GP individuals and, hence, the production of more simple solutions [26]. Consequently, a GP individual with a larger fitness value is considered to be more accurate in two-category discrimination.

2.3.2.5 Selection Procedure

The selection of individuals to produce successive generations plays an extremely important role in GP. There are various fitness-based selection methods, amongst which the tournament selection is recognised as the mainstream method for a GP selection procedure [22]. The tournament selection operates on subsets of individuals in a population. A randomly chosen number of individuals, defined by the tournament size, form a subset, where a selection competition is performed. Best individuals from the subsets are then passed to the next level, where the competition is repeated. The tournament selection allows to adjust the selection pressure [27], which is an objective measure to the characterise convergence rate of the selection, i.e. the smaller the tournament size, the lower the pressure [22].

2.4 Particle Swarm Optimisation

The standard particle swarm optimiser (SPSO) is a population-based algorithm that was invented by Kennedy and Eberhart [9], which was inspired by the social behaviour of animals such as fish schooling and bird flocking. Similar to other population-based algorithms, such as GAs, SPSO can not only solve a variety of difficult optimisation problems but also has shown a faster convergence rate than other EAs for some problems [10]. Another advantage of SPSO is that it has very few parameters to adjust, which makes it particularly easy to implement.

Angeline [28] pointed out that although SPSO may outperform other EAs in early iterations, its performance may not be competitive as the number of generations is increased. Recently, investigations have been undertaken to improve the performance of SPSO. Løvbjerg et al. [29] presented a hybrid PSO model with breeding and subpopulations. Kennedy and Mendes [30] studied the impacts of population structures to the search performance of SPSO. Other investigations on improving SPSO's performance were undertaken using the cluster analysis [31] and the fuzzy adaptive inertia weight [32]. SPSO has been used to tackle various engineering problems as presented in [33].

The foundation of SPSO is stemmed on the hypothesis that social sharing of information amongst conspecifics offers an evolutionary advantage [9], and the SPSO model is rooted on the following two factors [9]:

1. The autobiographical memory, which remembers the best previous position of each individual (P_i) in a swarm.
2. The publicised knowledge, which is the best solution (P_g) found currently by a population.

Therefore, the sharing of information amongst conspecifics is achieved by employing the publicly available information P_g, shown in Fig. 2.6. There is no information sharing amongst individuals except that P_g broadcasts the information to the other individuals. Therefore, a population may lose diversity and is more

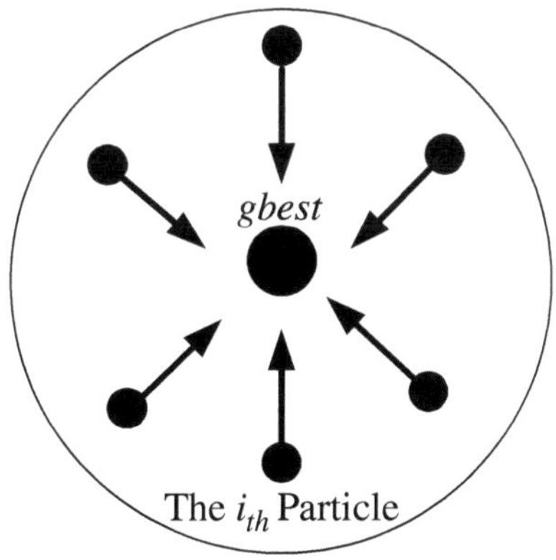

Fig. 2.6 Interaction between particles and the best particle *gbest*

likely to confine the search around local minima if committed too early in the search to the global best found so far.

To overcome this weakness, ideas from biology science have been borrowed to avoid early convergence and biologists have proposed four types of biological mechanisms that allow animals to aggregate into groups: passive aggregation, active aggregation, passive congregation and social congregation [34]. There are different information sharing mechanisms inside these forces. It is found that the passive congregation model is suitable to be incorporated in the SPSO model. Inspired by this observation, a hybrid model of PSO with passive congregation is presented in this book [11].

2.4.1 Standard Particle Swarm Optimisation

The population of SPSO is called a swarm and each individual is called a particle. For the ith particle at iteration k, it has the following two attributes.

1. A current position in an N-dimensional search space $X_i^k = (x_{i,1}^k, \ldots, x_{i,n}^k, \ldots, x_{i,N}^k)$, where $x_{i,n}^k \in [l_n, u_n]$, $1 \leq n \leq N$, l_n and u_n are the lower and upper bounds for the nth dimension, respectively.
2. A current velocity V_i^k

$$V_i^k = (v_{i,1}^k, \ldots, v_{i,n}^k, \ldots, v_{i,N}^k)$$

which is clamped to a maximum velocity

$$V_{\max}^k = (v_{\max,1}^k, \ldots, v_{\max,n}^k, \ldots, v_{\max,N}^k).$$

In each iteration, the swarm is updated by the following equations [9]:

$$V_i^{k+1} = \omega V_i^k + c_1 r_1 (P_i^k - X_i^k) + c_2 r_2 (P_g^k - X_i^k) \tag{2.6}$$

$$X_i^{k+1} = X_i^k + V_i^{k+1} \tag{2.7}$$

where P_i is the best previous position of the ith particle (also known as *pbest*) and P_g is the global best position amongst all the particles in the swarm (also known as *gbest*). They are given by the following equations:

$$P_i = \begin{cases} P_i & : \; f(X_i) \geq P_i \\ X_i & : \; f(X_i) < P_i \end{cases} \tag{2.8}$$

$$P_g \in \{P_0, P_1, \ldots, P_m\} | f(P_g) = \min(f(P_0), f(P_1), \ldots, f(P_m)) \tag{2.9}$$

where f is the objective function, m is the number of particles, r_1 and r_2 are the elements from two uniform random sequence on the interval $[0, 1] : r_1 \sim U(0, 1)$; $r_2 \sim U(0, 1)$ and ω an inertia weight which is typically chosen in the range of [0,1]. A larger inertia weight facilitates the global exploration and a smaller inertia weight tends to facilitate the local exploration to fine-tune the current search area [35]. Therefore, the inertia weight ω is critical for SPSO's convergence behaviour. A suitable value of ω usually provides a balance between global and local exploration abilities and consequently results in a better optimum solution. c_1 and c_2 are acceleration constants, which also control how far a particle moves in a single iteration. The maximum velocity $V_{\max}$ is set to be half of the length of the search space.

2.4.2 Particle Swarm Optimisation with Passive Congregation

It is mentioned that SPSO is inspired by social behaviours such as spatial order, more specially, aggregation such as bird flocking, fish schooling, or swarming of insects. Each of these cases has stable spatio-temporal integrities of a group of organisms: the group moves persistently as a whole without losing the shape and density.

For each of these groups, different biological forces are essential for preserving the group's integrity. Parrish and Hamner [34] proposed mathematical models of the spatial structure of animal groups to show how animals organise themselves. In these models, aggregation sometimes refers to a grouping of the organisms by non-social, external and physical forces. There are two types of aggregation: passive aggregation and active aggregation. Passive aggregation is a passive grouping by physical processes. One example of passive aggregation is the dense aggregation of plankton in open water, in which the plankton are not attracted actively to the aggregation but are transported passively there via physical forces such as water currents. Active aggregation is a grouping by attractive resources, such as food or space, with each member of the group recruited to a specific location actively. Congregation, which is different from aggregation, is a grouping by social forces, which is the source of attraction, in the group itself. Congregation can be classified into passive congregation and social congregation. Passive congregation is an attraction of an individual to other group members but where there is no display of social behaviour. Social congregations usually happen in a group where the members are related (sometimes highly related). A variety of inter-individual behaviours are displayed in social congregations, necessitating active information transfer [34]. For example, ants use antennal contacts to transfer information about an individual identity or a location of resources [36].

From the definitions above, the third part of Eq. 2.6: $c_2 r_2 (P_g^k - X_i^k)$ can be classified as either active aggregation or passive congregation. Since P_g is the best solution a swarm has found so far, which can be regarded as the place with most

Fig. 2.7 Search direction of the ith particle in SPSO

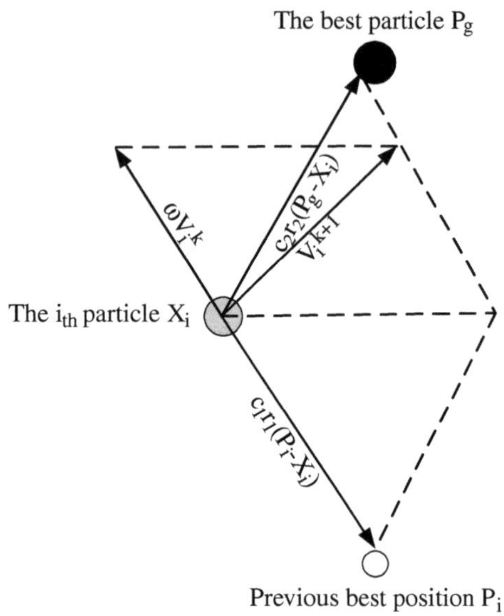

food, we argue that it is better to classify $c_2r_2(P_g^k - X_i^k)$ as active aggregation. From a biology point of view, the sharing of information amongst conspecifics is achieved by employing the publicly available information *gbest*. There is no information sharing amongst individuals except that *gbest* gives out the information to the other individuals. Therefore, for the ith particle, the search direction is only affected by 3 factors as shown in Fig. 2.7: the inertia velocity ωV_i^k, the best previous position *pbest*, and the position of global best particle *gbest*. The population is more likely to lose diversity and confine the search around local minima. From our experiment results, the performance of SPSO is not sufficiently good enough for high-dimensional and multi-model optimisation problems.

It has been discovered that in spatially well-defined congregations, such as fish schools, individuals may have low fidelity to a group because the congregations may be composed of individuals with little to no genetic relation to each other [37]. Schooling fish are generally considered as a "selfish herd" [38], in which each individual attempts to take the sweeping generalisation advantage from group living, independent of the fates of neighbours [39]. In these congregations, information may be transferred passively rather than actively [40]. Such asocial types of congregations can be referred to as passive congregation. As SPSO is inspired by fish schooling, it is, therefore, natural to ask if a passive congregation model can be employed to improve the performance of SPSO. Here, we do not consider other models such as passive aggregation, because SPSO is not aggregated passively via physical processes. Furthermore, social congregation usually happens when group fidelity is high, i.e. the chance of each individual meeting any of the others is high [41]. Social congregations frequently display a division of labour. In a social insect colony, such as an ant colony, large tasks are accomplished collectively by groups of specialised individuals, which is more efficient than performing sequentially by unspecialised individuals [42]. The concept of labour division can be employed by data clustering, sorting [43] and data analysis [44].

Group members in an aggregation can react without direct detection of incoming signals from an environment, because they can get necessary information from their neighbours [34]. Individuals need to monitor both environment and their immediate surroundings, such as the bearing and speed of their neighbours [34]. Therefore, each individual in an aggregation has a multitude of potential information from other group members that may minimise the chance of missed detection and incorrect interpretations [34]. Such information transfer can be employed in the model of passive congregation. Inspired by this perception, and to keep the model simple and uniform with SPSO, a hybrid PSO with passive congregation is proposed:

$$V_i^{k+1} = \omega V_i^k + c_1 r_1 (P_i^k - X_i^k) + c_2 r_2 (P_g^k - X_i^k) + c_3 r_3 (R_i^k - X_i^k) \tag{2.10}$$

$$X_i^{k+1} = X_i^k + V_i^{k+1} \tag{2.11}$$

where R_i is a particle randomly selected from the swarm, c_3 is the passive congregation coefficient and r_3 is a uniform random sequence in the range (0,1): $r_3 \sim U(0, 1)$. The interactions between individuals of PSOPC are shown in Fig. 2.8, and the pseudo code for implementing PSOPC is illustrated in Table 2.3.

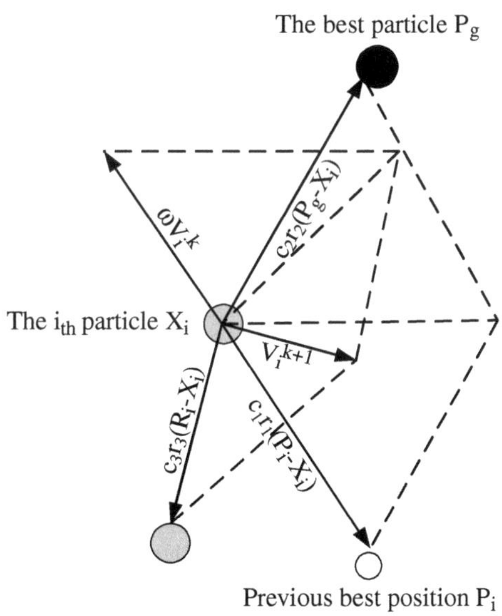

Fig. 2.8 Search direction of the *i*th particle in PSOPC

Table 2.3 Pseudo code for the implementation of PSOPC

Set $k = 0$;	
Randomly initialise positions;	
Randomly initialise velocities;	
WHILE (the termination conditions are not met)	
FOR (each particle i in the swarm)	
Check feasibility:	Check the feasibility of the current particle. If X_i^k is outside the feasible region, then reset X_i to the previous position X_i^{k-1};
Calculate fitness:	Calculate the fitness value $f(X_i)$ of the current particle;
Update *pbest*:	Compare the fitness value of *pbest* with $f(X_i)$. If $f(X_i)$ is better than the fitness value of *pbest*, then set *pbest* to the current position X_i;
Update *gbest*:	Find the global best position of the swarm. If $f(X_i)$ is better than the fitness value of *gbest*, then *gbest* is set to the position of the current particle X_i;
Update R_i:	Randomly select a particle from the swarm as R_i;
Update velocities:	Calculate velocities V_i using Eq. 2.10;
Update positions:	Calculate positions X_i using Eq. 2.11;
END FOR	
Set $k = k + 1$;	
END WHILE	

2.5 Summary

This chapter presents a brief introduction to evolutionary computation and its constitutive algorithms in order to provide a necessary background for the work discussed in later chapters. The basics of three EAs, i.e. GA, GP and PSO, are described, which are employed for identifying model parameters and evaluating fault features. First, the principles of GA are described, as well as the implementation procedures of an SGA. Then, the foundation of GP is presented, including the definition of terminals and functions, genetic operators, population initialisation and selection procedures. Finally, the standard PSO algorithm is introduced, followed by a description of an improved PSO algorithm.

References

1. Goldberg D (1989) Genetic algorithms in search, optimization, and machine learning. Addison-Wesley Publishing Company, Inc., USA
2. Coello CAC, Veldhuizen DAV, Lamont GB (2002) Evolutionary algorithms for solving multi-objective problems. Kluwer Academic Publishers, New York
3. Koza JR (1992) Genetic programming: on the programming of computers by means of natural selection. MIT Press, Cambridge

4. Fogel LJ (1994) Evolutionary programming in perspective: the top-down view. In: Zurada JM, Marks II RJ, Robinson CJ (eds) Computational intelligence: imitating life. IEEE Press, Piscataway
5. Rechenberg I (1994) Evolution strategy. In: Zurada JM, Marks II RJ, Robinson C (eds) Computational intelligence: imitating life. IEEE Press, Piscataway
6. Reeves CR, Rowe JE (2003) Genetic algorithms—principles and perspectives (a guide to GA theory). Kluwer Academic Publishers, London
7. Miettinen K (1999) Evolutionary algorithms in engineering and computer science. JohnWiley and Sons Inc., UK
8. Bäck T (1996) Evolution algorithms in theory and practice. Oxford University Press, NY
9. Kennedy J, Eberhart RC (1995) Particle swarm optimization. In: IEEE international conference on neural networks, vol 4. IEEE Press, NY, pp 1942–1948
10. Kennedy J, Eberhart RC (2001) Swarm intelligence. Morgan Kaufmann Publishers, San Francisco
11. He S, Wu QH, Wen JY, Saunders JR, Paton RC (2004) A particle swarm optimizer with passive congregation. BioSystems 78(1–3):135–147
12. Holland J (1975) Adaptation in natural and artificial systems. University of Michigan Press, USA
13. Wu QH, Cao YJ, Wen JY (1998) Optimal reactive power dispatch using an adaptive genetic algorithm. Int J Electr Power Energy Syst 20(8):563–569
14. Walters DC, Sheble GB (1993) Genetic algorithm solution of economic dispatch with valve point loading. IEEE Trans Power Syst 8(3):1325–1332
15. Wu QH, Ma JT (1995) Power system optimal reactive dispatch using evolutionary programming. IEEE Trans Power Syst 10(3):1243–1249
16. Iba K (1994) Reactive power optimization by genetic algorithm. IEEE Trans Power Syst 9(2):685–692
17. Lee KY, Bai X, Park YM (1995) Optimization method for reactive power planning using a genetic algorithm. IEEE Trans Power Syst 10(4):1843–1850
18. Ma JT, Wu QH (1995) Generator parameter identification using evolutionary programming. Int J Electr Power Energy Syst 17(6):417–423
19. Zhao Y, Edwards RM, Lee KY (1997) Hybrid feedforward and feedback controller design for nuclear steam generators over wide range operation using genetic algorithm. IEEE Trans Energy Conv 12(1):100–106
20. Michalewicz Z (1994) Genetic algorithms + data structures = evolution programs. AI series. Springer-Verlag, New York
21. Langdon WB (1998) Genetic programming and data structures: genetic programming + data structures = automatic programming. Kluwer Academic Publishers, Boston
22. Banzhaf W, Nordin P, Keller RE, Francone FD (1998) Genetic programming-an introduction: on the automatic evolution of computer program and its applications. Morgan Kaufmann Publishers Inc., San Francisco
23. Koza JR (1994) Genetic programming: automatic discovery of reusable programs. MIT Press, Cambridge
24. Guo H, Jack LB, Nandi AK (2005) Feature generation using genetic programming with application to fault classification. IEEE Trans Syst Man Cybernet B: Cybernet 35(1):89–99
25. Theodoridis S, Koutroumbas K (2003) Pattern recognition, 2nd edn. Academic Press, London
26. Zhang L, Jack LB, Nandi AK (2005) Fault detection using genetic programming. Mech Syst Signal Process 19:271–289
27. Bäck T, Fogel DB, Michalewicz Z (2000) Evolutionary computation 1: basic algorithms and operators. Institute of Physics Publishing Ltd., Bristol
28. Angeline P (1998) Evolutionary optimization versus particle swarm optimization: philosophy and performance difference. In: Proceedings of the evolutionary programming conference, San Diago, USA

29. Løvbjerg M, Rasmussen T, Krink K (2001) Hybrid particle swarm optimizer with breeding and subpopulations. In: Proceedings of the third genetic and evolutionary computation conference (GECCO-2001), vol 1. pp 469–476
30. Kennedy J, Mendes R (2002) Population structure and particle swarm performance. In: Proceedings of the 2002 congress on evolutionary computation CEC2002. IEEE Press, pp 1671–1676
31. Kennedy J (2000) Stereotyping: improving particle swarm performance with cluster analysis. In: Proceedings of the IEEE international conference on evolutionary computation. pp 1507–1512
32. Shi Y, Eberhart RC (2001) Fuzzy adaptive particle swarm optimization. In: Proceedings of the IEEE international conference on evolutionary computation. pp 101–106
33. Yoshida H, Kawata K, Fukuyama Y, Takayama S, Nakanishi Y (2000) A particle swarm optimization for reactive power and voltage control considering voltage security assessment. IEEE Trans Power Syst 15(4):1232–1239
34. Parrish JK, Hamner WM (1997) Animal groups in three dimensions. Cambridge University Press, Cambridge
35. Shi Y, Eberhart RC (1998) Parameter selection in particle swarm optimization. Evolutionary programming VII (1998). Lecture notes in computer science, vol 1447. Springer, NY, pp 591–600
36. Gordon DM, Paul RE, Thorpe K (1993) What is the function of encounter pattern in ant colonies? Anim Behav 45:1083–1100
37. Hilborn R (1991) Modelling the stability of fish schools: exchange of individual fish between schools of skipjack tuna (Katsuwonus pelamis). Canad J Fish Aqua Sci 48:1080–1091
38. Hamilton WD (1971) Geometry for the selfish herd. J Theor Biol 31:295–311
39. Pitcher TJ, Parrish JK (1993) Functions of shoaling behaviour in teleosts. Pitcher TJ (ed) Behaviour of teleost fishes. Chapman and Hall, London, pp 363–439
40. Magurran AE, Higham A (1988) Information transfer across fish shoals under predator threat. Ethology 78:153–158
41. Alexander RD (1974) The evolution of social behaviour. Annu Rev Ecol Syst 5:325–383
42. Bonabeau E, Dorigo M, Theraulaz G (1999) Swarm intelligence: from natural to artificial systems. Oxford University Press, USA
43. Deneubourg JL, Goss S, Franks N, Sendova-Franks A, Detrain C, Chretien L (1991) The dynamics of collective sorting: robot-like ant and ant-like robot. In: Proceedings of the first conference on simulation of adaptive behavior: from animals to animals. pp 356–365
44. Lumer E, Faieta B (1994) Diversity and adaptation in population of clustering ants. In: Proceedings of the third international conference on simulation of adaptive behavior: from animals to animals. pp 499–508

Chapter 3
Methodologies Dealing with Uncertainty

Abstract Uncertainties may arise in complex human thinking processes, which can become particularly challenging in the decision-making context for hard engineering problems with vague, imprecise and incomplete knowledge and information. As an important branch of computational intelligence, the logical approach can be employed to deal with such uncertainties. This chapter presents three mathematical theories, i.e. the Dempster–Shafer theory, the probability theory and the fuzzy logic (FL) theory, to handle different kinds of uncertainties. The FL theory can deal with the imprecision (or vagueness) of defined knowledge, whilst the Dempster–Shafer theory provides two measures (support and plausibility) for formulating a mechanism to represent "ignorance". As a probabilistic technique, Bayesian networks (BNs) are introduced as graphical representations of uncertain knowledge. In later chapters, the three methodologies are employed to tackle uncertainties arising from complicated condition assessment procedures for detecting transformer faults.

3.1 The Logical Approach of Computational Intelligence

The logical approach is an important branch of CI, which is initially based upon the use of the if–then logical constructions to develop a theorem proving programs. These programs analyse the knowledge about properties and relationships amongst objects to solve a problem (derive a procedure, conclusion, etc.), which would normally require a human expert. Such CI programs are called EPSs, which are able to make choices analysing information acquired from a variety of knowledge sources, e.g. human experts, databases and so forth [1]. To make expert systems more efficient in dealing with vague, imprecise and uncertain knowledge and information, the FL concept was introduced [2], which used fuzzy values to encode human knowledge in a form that reflected accurately experts' understanding of difficult, complex problems and, thus, capture human reasoning

W. H. Tang and Q. H. Wu, *Condition Monitoring and Assessment of Power Transformers Using Computational Intelligence*, Power Systems,
DOI: 10.1007/978-0-85729-052-6_3,

and decision making [3]. Alternatively, a probability theory can also be employed to express uncertain information, upon which a variety of evidence combination and decision making methods have been successfully developed, utilising different concepts of probabilistic reasoning. Amongst the most applied methods are the Bayesian [4] theory and the Dempster–Shafer theory [5].

There are some subtle differences amongst the previously mentioned three mathematical theories in handling different kinds of uncertainties, i.e. the Dempster–Shafer theory, the probability theory and the FL theory. In simple terms, a probability represents the chance of an instance belonging to a particular concept class given the current knowledge, whilst the FL is treated as a mechanism used to deal with the imprecision (or vagueness) of the defined knowledge (i.e. the defined concept does not have a certain extension in semantics, and the FL can be used to specify how well an object satisfies such a vague description). The Dempster–Shafer theory, on the other hand, is a mechanism for handling "ignorance", which provides two measures (support and plausibility) for representing beliefs about propositions and works well in simple rule-based systems due to its high computational complexity. In this chapter, the three mathematical theories are introduced briefly, which are employed in later chapters to deal with uncertainties arising from complicated condition assessment procedures for power transformers.

3.2 Evidential Reasoning

It is always desirable to find an information aggregation method, which could be used to tackle MADM problems in engineering. In this section, an ER approach is presented, which deals with uncertain decision knowledge in MADM problems on the basis of a firm mathematical foundation. The ER approach was first established in [6] and generalised in [7] to provide a multi-attribute evaluation framework for overall decision making by aggregating subjective judgement on constitutive attributes for an MADM problem. Regarding transformer condition assessment, the ER has been employed to develop a formalised evaluation framework to integrate various diagnoses obtained with traditional diagnostics methods for transformer DGA [8, 9]. The kernel of the approach is an ER algorithm, developed upon the evidence combination rules of the Dempster–Shafer theory, which is used to aggregate attributes of a multi-level structure [7]. Recently, the algorithm has been revised in [10] with an updated weight normalisation scheme and a simple probability assignment strategy. In this section, the basics of the original ER algorithm and the revised ER algorithm are introduced in detail, as well as the ranking processes used by an ER approach.

3.2.1 The Original Evidential Reasoning Algorithm

A hybrid MADA problem may be expressed using the following formula [5, 6, 11].

$$\text{maximise}_{a\in\Omega} \quad y(a) = [y_1(a), \ldots, y_k(a), \ldots, y_{k_1+k_2}(a)], \tag{3.1}$$

where, Ω is a discrete set of alternatives ($\Omega = [a_1,\ldots,a_r]$, $r = 1,\ldots,l$), $y(a)$ the overall evaluation of alternative a, $y_k(a)$ the evaluation of the kth attribute of $y(a)$ and k_1 and k_2 the numbers of quantitative and qualitative attributes of each alternative, respectively. For transformer condition assessment, alternatives represent a group of transformers. An extended decision matrix for qualitative and quantitative attributes is presented in Table 3.1, where y_{ij} is a numerical value of y_j at a_i (i=1,…,l; j=1,…,k_1) and SJ_{ij} the subjective judgements with uncertainties for evaluating the states of y_{k_1+j} at a_i (i=1,…,l; j=1,…,k_2). The objective is to rank these alternatives or to select the best compromise alternative, with both quantitative and qualitative attributes being satisfied as much as possible.

An evaluation analysis model concerning only qualitative judgements is presented in Fig. 3.1, which defines a typical evaluation structure for an MADA problem. There are three levels in the evaluation model, i.e. a factor level, an evaluation grade level and an attribute level. In the attribute level of the evaluation model, the state of an attribute at each alternative a is required to be evaluated. In the evaluation grade level, a simple evaluation method is used to define a few evaluation grades, so that the state of an attribute of an alternative could be evaluated to one of the predefined grades. These grades may be quantified using certain scales. In the evaluation grade level, H_n is called an evaluation grade (n=1,…,N). A set of evaluation grades for evaluating an attribute y_k is denoted as

$$H = \{H_1, H_2, \ldots, H_n, \ldots, H_N\}, \tag{3.2}$$

where N is the number of evaluation grades. H_n represents a grade to which the state of y_k may be evaluated. H_1 and H_N are set to be the worst and the best grades, respectively, and H_{n+1} is supposed to be preferred to H_n.

The qualitative evaluation is difficult to give, as it is subjective and sometimes incomplete. In order to quantify these evaluation grades and eventually to quantify subjective judgements possessing uncertainties, the concept of preference degree is introduced. A preference degree takes values from a closed interval [−1, 1] ([worst, best]), which may be called a preference degree space. The set of evaluation grades may, thus, be quantified by [6]

Table 3.1 An extended decision matrix including both quantitative and qualitative attributes

Alternative	Quantitative attributes (y_k)				Qualitative attributes (y_k)			
(a_r)	y_1	y_2	…	y_{k_1}	y_{k_1+1}	y_{k_1+2}	…	$y_{k_1+k_2}$
a_1	y_{11}	y_{12}	…	y_{1k_1}	SJ_{11}	SJ_{12}	…	SJ_{1k_2}
a_2	y_{21}	y_{22}	…	y_{2k_1}	SJ_{21}	SJ_{22}	…	SJ_{2k_2}
…	…	…	…	…	…	…	…	…
a_l	y_{l1}	y_{l2}	…	y_{lk_1}	SJ_{l1}	SJ_{l2}	…	SJ_{lk_2}

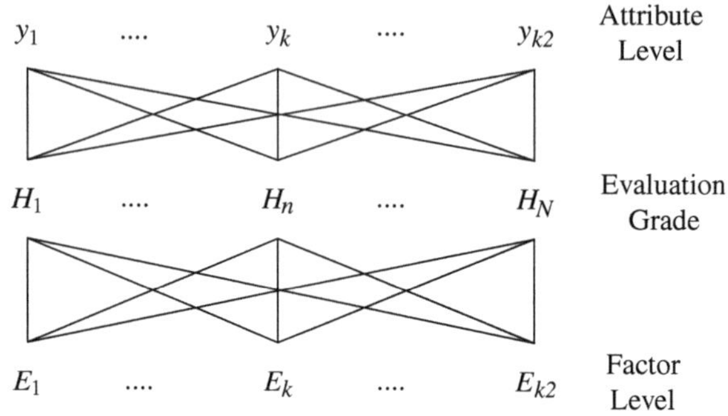

Fig. 3.1 An evaluation analysis model dealing with subjective judgements

$$p\{H\} = [p(H_1), \ldots, p(H_n), \ldots, p(H_N)]^{\mathrm{T}}, \tag{3.3}$$

where $p\{H\}$ is the numerical scale of H_n and satisfies the following basic conditions:

$$p(H_1) = -1, \quad p(H_N) = 1,$$

$$p(H_{n+1}) > p(H_n), \quad n = 1, \ldots, N-1. \tag{3.4}$$

In the factor level, E_k represents a set of factors, which is associated with the evaluation of attribute $y_k(a)$ and denoted by

$$E_k = \{e_k^1, e_k^2, \ldots e_k^{L_k}\} \quad k = k_1 + 1, \ldots, k_1 + k_2, \tag{3.5}$$

where e_k^i ($i = 1, \ldots, L_k$) are factors influencing the evaluation of $y_k(a)$. The state of e_k^i can be evaluated directly at an alternative a, i.e. $e_k^i = e_k^i(a)$. Hence, the preference degree for the state of an attribute $y_k(a)$ through the direct evaluations of the relevant factors e_k^i can be generated using the Dempster–Shafer theory, which is introduced as below.

3.2.1.1 The Outline of the Dempster–Shafer Theory

The Dempster–Shafer theory was first set forth by Dempster in the 1960s and subsequently extended by Shafer [5, 11].

The Dempster–Shafer theory defines a frame of discernment, denoted by Θ. A basic hypothesis (singleton) in Θ is defined as H_s, i.e. $H_s \subseteq \Theta$. The hypotheses in Θ are assumed mutually exclusive and exhaustive. A subset of the hypotheses in Θ gives rise to a new hypothesis, which is equivalent to the disjunction of the hypotheses in the subset. Each hypothesis in Θ corresponds to a one-element subset (called a singleton). By considering all possible subsets of Θ, denoted as 2^{Θ}, the set of hypotheses to which belief can be allotted is enlarged. Henceforth, the term hypothesis is used in this enlarged sense to denote any subset of the original hypotheses in Θ. Note that a set of size n has 2^n subsets.

The Dempster–Shafer theory uses a number in the range [0, 1] to indicate belief in a hypothesis given a piece of evidence. This number is the degree to which the evidence supports the hypothesis. Recall that evidence against a hypothesis is regarded as evidence for the negation of the hypothesis. The impact of each distinct piece of evidence on the subsets of Θ is represented by a function called a basic probability assignment (BPA). A BPA is a generalisation of traditional probability density functions, which assigns a number in the range [0, 1] to every singleton of Θ so that the numbers sum to 1. Using 2^{Θ}, the enlarged domain of all subsets of Θ, a BPA, denoted as m, assigns a number in [0, 1] to every subset of Θ such that the numbers sum to 1. Thus, m allows an assignment of a quantity of belief to every element, not just to those elements on the bottom row in Fig. 3.1, as is the case for a probability density function.

The quantity $m(A)$ is a measure of that portion of the total belief committed exactly to A, where A is an element of 2^{Θ} and the total belief is 1. The portion of belief cannot be further subdivided amongst the subsets of A and does not include portions of belief committed to subsets of A. The quantity $m(\Theta)$ is a measure of that portion of the total belief that remains unassigned after commitment of belief to various proper subsets of Θ. If $m(A) = s$ and s assigns no belief to other subsets of Θ, then $m(\Theta) = 1 - s$. Thus, the remaining belief is assigned to Θ and not to the negation of the hypothesis, as would be required in the traditional Bayesian model [11].

If there exists two pieces of evidence in Θ, they provide two BPAs to a subset Ψ of Θ, i.e. $m_1(\Psi)$ and $m_2(\Psi)$. The problem is to obtain a combined probability assignment $m_{12}(\Psi) = m_1(\Psi) \oplus m_2(\Psi)$. The Dempster–Shafer theory provides an evidence combination rule defined below:

$$m_{12}(\Phi) = 0, \quad m_{12}(\Psi) = \sum_{A \cap B = \Psi} \frac{m_1(A)m_2(B)}{1 - K},$$

$$K = \sum_{A \cap B = \Phi} m_1(A)m_2(B).$$

In the rule, $m_{12}(\Psi)$ for hypothesis $\Psi(\subseteq\Theta)$ is computed from m_1 and m_2 by adding all products of the form $m_1(A)m_2(B)$ where A and B are selected from the subsets of Θ in all possible ways such that their intersection is Ψ. K reflects the conflicting situations where both $m_1(A)$ and $m_1(B)$ are not zero, but the intersection $A \cap B$ is empty. The commutativity of multiplication in the rule ensures that the rule yields the same value regardless of the order in which the two pieces of evidence are combined.

3.2.1.2 An ER Framework for an Evaluation Analysis Model

In the previously presented evaluation analysis model, an evaluation grade H_n may be considered as a basic hypothesis in the Dempster–Shafer theory and a factor e_k^i

is regarded as a piece of evidence. All the evaluation grades in H are defined as distinct grades, and also must cover all possible grades. Then, the frame of discernment may be defined by

$$\Theta = H = \{H_1, \ldots, H_n, \ldots, H_N\}. \tag{3.6}$$

Let $m(H_n/e_k^i(a))$ expresses a BPA in which e_k^i supports a hypothesis that the state of y_k at an alternative a is confirmed to H_n. In this study, a BPA is derived from a confidence degree [6]. Let $\beta(e_k^i(a))$ be a confidence degree to which a decision maker considers that the state of e_k^i at an alternative a is confirmed. If there is only one factor e_k^i in E_k, $m(H_n/e_k^i(a))$ should be equal to $\beta(e_k^i(a))$. If there are multiple factors in E_k, the confidence degree $\beta(e_k^i(a))$ depends upon a weighted confidence degree λ_k^i. λ_k^i is the normalised relative weight of e_k^i in E_k and $\lambda_k = [\lambda_k^1, \ldots, \lambda_k^{L_k}]^T$. Then $m(H_n/e_k^i(a))$ is determined by

$$m(H_n/e_k^i(a)) = \lambda_k^i \beta(e_k^i(a)). \tag{3.7}$$

A priority coefficient α_k may also be used in the above equation to represent the importance of the role of the most important factor for evaluating y_k. In the factor level, a set of factors $E_k = [e_k^1, e_k^2, \ldots, e_k^{L_k}]$ is classified into $N - 1$ subjects S_n defined by

$$S_n = \{e_{n,n+1}^1, \ldots, e_{n,n+1}^i, \ldots, e_{n,n+1}^{R_n}\}, \quad (n = 1, \ldots, N-1;\ i = 1, \ldots, R_n), \tag{3.8}$$

where $e_{n,n+1}^i$ is a factor in E_k, the state of which is confirmed to H_n and/or to H_{n+1}, and $L_k = R_1 + R_2 + \cdots + R_{N-1}$ (R_n is the number of involved factors). In this way, the BPAs can be generated from uncertain subjective judgements. The overall probability assignment can then be obtained by combining all the BPAs using the operational algorithms as presented in the following subsection.

3.2.1.3 Implementation of Dempster–Shafer Combination Rules

In the preceding section, a set of factors has been classified into $N - 1$ subsets denoted by S_n (n=1,…,$N - 1$). In S_n there are R_n factors as defined by Eq. 3.8, the states of which may be confirmed to H_n and/or H_{n+1}. In this subsection, an algorithm is formulated to generate local probability assignments to H_n and H_{n+1} by combining these R_n factors using partial combination rules first [6].

Suppose the basic probability assignments $m_n^{n,i}$ and $m_{n+1}^{n,i}$ (i=1,…,R_n) are obtained using Eq. 3.8, then $m_\Theta^{n,i} = 1 - (m_n^{n,i} + m_{n+1}^{n,i})$ (i=1,…,R_n). All these basic probability assignments to H_n, H_{n+1} and Θ with respect to $e_{n,n+1}^i$ (i=1,…,R_n; n=1,…,$N-$ 1) may then be expressed by the following basic probability assignment matrix $\mathbf{M}^n$:

$$\mathbf{M}^n = \begin{bmatrix} m_n^{n,1} & m_{n+1}^{n,1} & m_\Theta^{n,1} \\ m_n^{n,2} & m_{n+1}^{n,2} & m_\Theta^{n,2} \\ \cdots & \cdots & \cdots \\ m_n^{n,R_n} & m_{n+1}^{n,R_n} & m_\Theta^{n,R_n} \end{bmatrix} \begin{matrix} \{e_{n,n+1}^1\} \\ \{e_{n,n+1}^2\} \\ \cdots \\ \{e_{n,n+1}^{Rn}\} \end{matrix}, \quad (n = 1, \ldots, N-1).$$

If $R_n = 0$, then $m_n^{n,i} = 0$, $m_{n+1}^{n,i} = 0$ and $m_\Theta^{n,i} = 1$. Let $m_n^{I(1)} = m_n^{n,1}$, $m_{n+1}^{I(1)} = m_{n+1}^{n,1}$ and $m_\Theta^{I(1)} = m_\Theta^{n,1}$, then it is logical by combining $e_{n,n+1}^{I(r+1)} = \{e_{n,n+1}^1, \ldots, e_{n,n+1}^{r+1}\}$ to obtain the following recursive formulae:

$$\{H_n\} : m_n^{I(r+1)} = K^{I(r+1)}(m_n^{I(r)} m_n^{n,r+1} + m_n^{I(r)} m_\Theta^{n,r+1} + m_\Theta^{I(r)} m_n^{n,r+1}), \tag{3.9}$$

$$\{H_{n+1}\} : m_{n+1}^{I(r+1)} = K^{I(r+1)}(m_{n+1}^{I(r)} m_{n+1}^{n,r+1} + m_{n+1}^{I(r)} m_\Theta^{n,r+1} + m_\Theta^{I(r)} m_{n+1}^{n,r+1}), \tag{3.10}$$

$$\{\Theta\} : m_\Theta^{I(r+1)} = K^{I(r+1)} m_\Theta^{I(r)} m_\Theta^{n,r+1}, \tag{3.11}$$

where

$$\begin{aligned} K^{I(r+1)} &= [1 - (m_n^{I(r)} m_{n+1}^{n,r+1} + m_{n+1}^{I(r)} m_n^{n,r+1})]^{-1}, \\ r &= 1, \ldots, R_n - 1; \ n = 1, \ldots, N-1. \end{aligned} \tag{3.12}$$

The local probability assignments to H_n, H_{n+1} and Θ with respect to the subset of factors S_n can be represented as $m_n^{I(R_n)}$, $m_{n+1}^{I(R_n)}$ and $m_\Theta^{I(R_n)}$. To represent the results of the partial combination of all subsets of factors, the following matrix is defined, called the local probability assignment matrix $\mathbf{M}$:

$$\mathbf{M} = \begin{bmatrix} m_1^{I(R_1)} & m_2^{I(R_1)} & m_\Theta^{I(R_1)} \\ \cdots & \cdots & \cdots \\ m_n^{I(R_n)} & m_{n+1}^{I(R_n)} & m_\Theta^{I(R_n)} \\ \cdots & \cdots & \cdots \\ m_{N-1}^{I(R_{N-1})} & m_N^{I(R_{N-1})} & m_\Theta^{I(R_{N-1})} \end{bmatrix} \begin{matrix} \{e_{1,2}^{I(R_1)}\} \\ \cdots \\ \{e_{n,n+1}^{I(R_n)}\} \\ \cdots \\ \{e_{N-1,N}^{I(R_{N-1})}\} \end{matrix}. \tag{3.13}$$

After the partial combination procedure, the subset of factors S_n may be regarded as an aggregated factor and $m_n^{I(R_n)}$ as a new basic probability assignment to the hypothesis H_n, confirmed by S_n. The problem is then to combine all these integrated factors in order to obtain the overall probability assignments to all subsets Ψ of Θ, including the singletons H_n (n=1,…,N). Let $b_1^{C(1)} = m_1^{I(R_1)}$, $b_2^{C(1)} = m_2^{I(R_1)}$ and $b_\Theta^{C(1)} = m_\Theta^{I(R_1)}$, then we can combine $e_{1,j+2}^{C(j+1)} = \{e_{1,2}^{I(R_1)}, \ldots, e_{j+1,j+2}^{I(R_{j+1})}\}$ to obtain the following recursive algorithm [6]:

$$\{H_1\} : b_1^{C(j+1)} = K^{C(j+1)} b_1^{C(j)} m_\Theta^{I(R_{j+1})}, \tag{3.14}$$

$$\cdots$$

$$\{H_j\} : b_j^{C(j+1)} = K^{C(j+1)} b_j^{C(j)} m_\Theta^{I(R_{j+1})}, \tag{3.15}$$

$$\{H_{j+1}\}: b_{j+1}^{C(j+1)} = K^{C(j+1)}\Big(b_{j+1}^{C(j)} m_{j+1}^{I(R_{j+1})} + b_{j+1}^{C(j)} m_{\Theta}^{I(R_{j+1})} + b_{\Theta}^{C(j)} m_{j+1}^{I(R_{j+1})}\Big), \tag{3.16}$$

$$\{H_{j+2}\}: b_{j+2}^{C(j+1)} = K^{C(j+1)} b_{\Theta}^{C(j)} m_{j+2}^{I(R_{j+1})}, \tag{3.17}$$

$$\{\Theta\}: b_{\Theta}^{C(j+1)} = K^{C(j+1)} b_{\Theta}^{C(j)} m_{\Theta}^{I(R_{j+1})}, \tag{3.18}$$

where

$$K^{C(j+1)} = \left[1 - \left(\sum_{t=1}^{j} b_t^{C(j)}\left(m_{j+1}^{I(R_{j+1})} + m_{j+2}^{I(R_{j+1})}\right) + b_{j+1}^{C(j)} m_{j+2}^{I(R_{j+1})}\right)\right]^{-1} \tag{3.19}$$

$$j = 1, \ldots, N-2.$$

When $j = N - 2$, the overall probability assignments are generated, which can be expressed by the following vector and called the overall probability assignment vector:

$$G = [b_1^{C(N-1)}, \ldots, b_n^{C(N-1)}, \ldots, b_N^{C(N-1)}, b_{\Theta}^{C(N-1)}]^{\mathrm{T}}. \tag{3.20}$$

To sum up, G is obtained by combining $e_{1,N}^{C(N\text{-}1)}$ whilst

$$\begin{aligned} e_{1,N}^{C(N-1)} &= \{e_{1,2}^{I(R_1)}, \ldots, e_{n,n+1}^{I(R_n)}, \ldots, e_{N-1,N}^{I(R_{N-1})}\} \\ &= \{S_1, S_2, \ldots, S_n, \ldots, S_{N-1}\} \\ &= \{e_k^1, e_k^2, \ldots, e_k^{L_k}\} = E_k. \end{aligned} \tag{3.21}$$

It can also be noticed that $b_n^{C(N\text{-}1)}$ is the overall probability assignment to which H_n is confirmed by all factors e_k^i (i=1,…,L_k).

3.2.1.4 Construction of an Evaluation Matrix

The ER approach introduced above is actually employed to transform uncertain subjective judgements about the state of a qualitative attribute y_k at an alternative a_r into the preference degree $p_{rk} = p(y_k(a_r))$ for all $k = k_1 + 1, \ldots, k_1 + k_2$; r=1,…,l. In this way, all qualitative attributes are evaluated and quantified using numerical values in the interval [−1, 1].

The values of quantitative attributes which are generally incommensurate may also be transformed into the preference degree space using the following formulae [6]:

$$p_{rk} = p(y_{rk}) = \frac{2(y_{rk} - y_k^{\min})}{y_k^{\max} - y_k^{\min}} - 1, \quad k = 1, \ldots, k_1;\ r = 1, \ldots, l. \tag{3.22}$$

For benefit attributes

Table 3.2 The evaluation matrix

Preference degrees	$p(y_1)$	$\dots$	$p(y_{k_1})$	$p(y_{k_1+1})$	$\dots$	$p(y_{k_1+k_2})$
a_1	p_{11}	$\dots$	p_{1k_1}	p_{1k_1+1}	$\dots$	$p_{1k_1+k_2}$
a_2	p_{21}	$\dots$	p_{2k_1}	p_{2k_1+1}	$\dots$	$p_{2k_1+k_2}$
$\dots$	$\dots$	$\dots$	$\dots$	$\dots$	$\dots$	$\dots$
a_l	p_{l1}	$\dots$	p_{lk_1}	p_{lk_1+1}	$\dots$	$p_{lk_1+k_2}$

$$p_{rk} = p(y_{rk}) = \frac{2(y^{\max} - y_{rk})}{y_k^{\max} - y_k^{\min}} - 1, \quad k = 1, \dots, k_1;\ r = 1, \dots, l. \tag{3.23}$$

For cost attributes

$$\begin{aligned} y_k^{\max} &= \max\{y_{1k}, \dots, y_{lk}\}, \\ y_k^{\min} &= \min\{y_{1k}, \dots, y_{lk}\}. \end{aligned} \tag{3.24}$$

The transformed attribute y_k may be denoted by a preference function $p(y_k)$. Thus, the original extended decision matrix defined in Table 3.1 is transformed into an ordinary decision matrix as shown in Table 3.2, in which the states of all attributes, either quantitative or qualitative, are represented in the preference degree space. The alternative may then be ranked based on the developed evaluation matrix.

3.2.2 The Revised Evidential Reasoning Algorithm

The original ER algorithm has been developed upon an ER assessment framework as discussed in Sect. 3.2.1.2, which is formulated following four synthesis axioms, i.e. the basic synthesis theorem, the consensus synthesis theorem, the complete synthesis theorem and the incomplete synthesis theorem [10]. In [10] the original ER approach has been revised with an updated weight normalisation scheme and a simple probability assignment strategy that satisfies all the above four axioms as introduced below.

3.2.2.1 Definition of an Evidential Reasoning Evaluation Framework

Consider a three-level hierarchy of attributes, as shown in Fig. 3.1, with a general attribute y at the top level and a number of basic attributes at the bottom level. A set of basic attributes is defined as follows:

$$E = \{e_1, e_2, \dots, e_i, \dots, e_L\} \quad i = 1, \dots, L. \tag{3.25}$$

Each ith attribute e_i is assigned with a corresponding normalised weight ω_i ($0 \leq \omega_i \leq 1$), representing the relative importance of the attribute in an evaluation process. Thus, a set of weights is defined as

$$\omega = \{\omega_1, \omega_2, \ldots, \omega_i, \ldots, \omega_L\}. \tag{3.26}$$

The weights must satisfy the following condition:

$$\sum_{i=1}^{L} \omega_i = 1. \tag{3.27}$$

The state of an attribute is required to be assessed using a set of predefined evaluation grades (hypotheses)

$$H = \{H_1, H_2, \ldots, H_n, \ldots, H_N\} \quad n = 1, \ldots, N, \tag{3.28}$$

where N is the number of evaluation grades.

The generated assessment $S(e_i)$ for attribute e_i may be represented as the following distribution of degree of beliefs with regard to different hypotheses:

$$S(e_i) = \{(H_n, \beta_{n,i}),\ n = 1, \ldots, N\} \quad i = 1, \ldots, L, \tag{3.29}$$

which means that attribute e_i is assessed to grade H_n with the degree of belief of $\beta_{n,i}$ ($\beta_{n,i} \geq 0$ and $\sum_{n=1}^{N} \beta_{n,i} \leq 1$). The assessment $S(e_i)$ is complete if $\sum_{n=1}^{N} \beta_{n,i} = 1$ and incomplete if $\sum_{n=1}^{N} \beta_{n,i} < 1$. The case when $\sum_{n=1}^{N} \beta_{n,i} = 0$ (or $\beta_{n,i} = 0$ for all n=1,…,N) denotes a complete lack of information on e_i.

Let β_n be a degree of belief to which the general attribute y is assessed to grade H_n. To calculate β_n it is necessary to aggregate the assessments for all the associated basic attributes e_i given in the form of Eq. 3.29. The revised ER algorithm discussed as below can be used for this purpose, which is simpler than the original one introduced previously.

3.2.2.2 The Updated Scheme for Weight Normalisation and Basic Probability Assignment

In this subsection the revised ER algorithm is summarised briefly. The more detailed theory and discussion are presented extensively in [7, 10].

At first, degrees of belief are converted into basic probability masses. A basic probability mass $m_{n,i}$ represents the degree to which the ith basic attribute e_i supports the assessment of the general attribute y with the nth grade (hypothesis) H_n. It is calculated as follows:

$$m_{n,i} = \omega_i \beta_{n,i} \quad n = 1, \ldots, N. \tag{3.30}$$

The remaining probability mass $m_{H,i}$, unassigned to any individual grade, is decomposed into two parts $\overline{m}_{H,i}$ and $\tilde{m}_{H,i}$ such as

$$\overline{m}_{H,i} = 1 - \omega_i \quad \text{and} \quad \tilde{m}_{H,i} = \omega_i \left(1 - \sum_{n=1}^{N} \beta_{n,i}\right) \tag{3.31}$$

with $m_{H,i} = \overline{m}_{H,i} + \tilde{m}_{H,i}$.

Suppose $m_{n,I(i)}$, n=1,...,N, is the combined probability mass calculated by aggregating the first i assessments. $\overline{m}_{H,I(i)}$ and $\tilde{m}_{H,I(i)}$ are the remaining probability masses unassigned to any individual grade after the aggregation of the first i assessments. Assuming

$$m_{n,I(1)} = m_{n,1}, \quad \overline{m}_{H,I(1)} = \overline{m}_{H,1} \quad \text{and} \quad \tilde{m}_{H,I(1)} = \tilde{m}_{H,1},$$

the following recursive expressions are developed to combine the first i assessments with the $(i + 1)$th assessment:

$$\{H_n\} : \begin{array}{l} m_{n,I(i+1)} = K_{I(i+1)}\left[m_{n,I(i)}m_{n,i+1} + m_{H,I(i)}m_{n,i+1} + m_{n,I(i)}m_{H,i+1}\right], \\ m_{H,I(i)} \;\; = \tilde{m}_{H,I(i)} + \overline{m}_{H,I(i)} \quad n = 1, \ldots, N, \end{array} \tag{3.32}$$

$$\{H\} : \tilde{m}_{H,I(i+1)} = K_{I(i+1)}\left[\tilde{m}_{H,I(i)}\tilde{m}_{H,i+1} + \overline{m}_{H,I(i)}\tilde{m}_{H,i+1} + \tilde{m}_{H,I(i)}\overline{m}_{H,i+1}\right], \tag{3.33}$$

$$\{H\} : \overline{m}_{H,I(i+1)} = K_{I(i+1)}\left[\overline{m}_{H,I(i)}\overline{m}_{H,i+1}\right], \tag{3.34}$$

where $K_{I(i+1)}$ is defined as below for i = 1,..., $L - 1$,

$$K_{I(i+1)} = \left[1 - \sum_{t=1}^{N} \sum_{\substack{j=1 \\ j \neq t}}^{N} m_{t,I(i)} m_{j,i+1}\right]^{-1}. \tag{3.35}$$

After the aggregation of all the L basic attributes from the set E, β_n, corresponding to H_n, n=1,...,N, and the unassigned degree of belief β_H, representing the incompleteness of the overall assessment, are calculated using the following normalisation formulae:

$$\{H_n\} : \beta_n = \frac{m_{n,I(L)}}{1 - \overline{m}_{H,I(L)}} \quad n = 1, \ldots, N, \tag{3.36}$$

$$\{H\} : \beta_H = \frac{\tilde{m}_{H,I(L)}}{1 - \overline{m}_{H,I(L)}}. \tag{3.37}$$

Note that $\sum_{n=1}^{N} \beta_n + \beta_H = 1$.

Thus, the overall assessment for the general attribute y can be represented by the following distribution of degree of beliefs with regard to different hypotheses similar to Eq. 3.29:

$$S(y) = S(e_1 \oplus \cdots \oplus e_i \oplus \cdots \oplus e_L) = \{(H_n, \beta_n),\ n = 1, \ldots, N\}, \tag{3.38}$$

where $\oplus$ denotes the aggregation of two attributes.

3.3 Fuzzy Logic

3.3.1 Foundation of Fuzzy Logic

Fuzzy logic is based upon the fuzzy set theory, which was formalised by Professor Lofti Zadeh at the University of California in 1965 [2]. So what is a fuzzy set? A fuzzy set is a set without a crisp and clearly defined boundary. It contains elements with only a partial degree of membership. FL provides a means of calculating intermediate values between absolute true and absolute false with resulting values ranging between 0 and 1. In FL, a membership function is employed to calculate the degree to which an item is a member. Basically, a membership function is a curve that defines how each point in an input space is mapped to a membership value (or degree of membership) between 0 and 1. Typical membership functions include piece-wise linear functions, the sigmoid curve, the Gaussian distribution function and the quadratic and cubic polynomial curves. There are a number of logic operators to enable the fuzzy inference, e.g. the fuzzy intersection or conjunction (AND), fuzzy union or disjunction (OR) and fuzzy complement (NOT). If–then rule statements are used to formulate conditional statements that comprise FL. A typical implementation of FL includes: fuzzify inputs, apply fuzzy operators, apply an implication method, aggregate all outputs and defuzzify to make a decision.

3.3.2 An Example of a Fuzzy Logic System

Perhaps the best way to explain the concept of a fuzzy set is through an example. From classical mathematics we are familiar with the notion of a crisp set, e.g. if we consider a set X of real car speeds from 0 to 100 mph, which we call the universe of discourse, then a subset $S = [71, 100]$ defines all of the speeds that are classified as being over the U.K. highway speed limit. This function can be demonstrated clearly by using a function $f(S)$, where $f(S) = 1$ when an element of X belongs to S and 0 at all other times. Figure 3.2 shows the function $f(S)$.

The above definition is fine for a discrete problem such as a speed limit, but how would we classify a speed as slow, average or fast? If we use the same range of speeds X and now define 3 subsets as slow = [0, 30], average = [31, 60] and fast [61, 100], then using similar functions as before we would have a situation shown in Fig. 3.3. It is obvious that this is not an ideal representation since an increase of 1 mph cannot reasonably reclassify a car's speed from slow to average or from average to fast. In the real world the transition of a car's speed from slow to average is a gradual and continuous change with areas of uncertainty as to which classification should be made. A fuzzy set allows for this uncertainty, "fuzziness", by allocating each element a grade of membership, in the interval [0, 1], to each possible set (fuzzy set). Figure 3.4 illustrates one possible

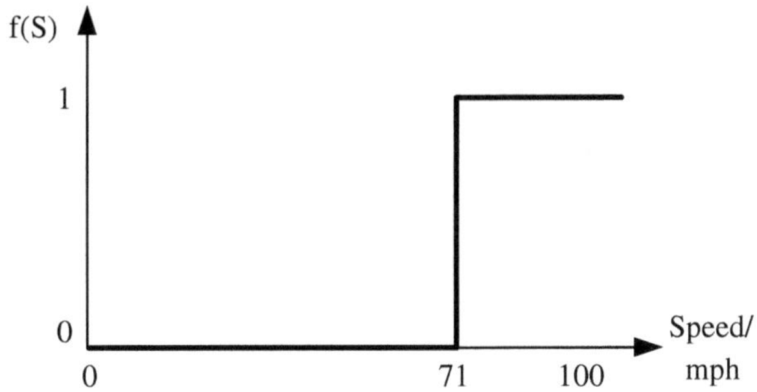

Fig. 3.2 Function describing membership of elements in *X* to subset *S*

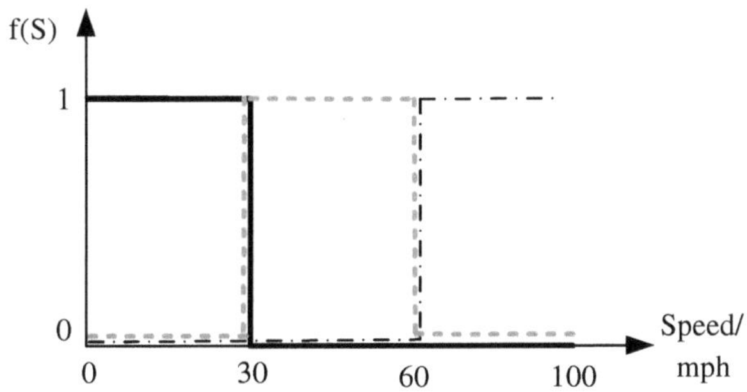

Fig. 3.3 Function describing membership of elements in *X* to the 3 subsets slow, average and fast

implementation of this idea. From Fig. 3.4, it is seen that now the speed 30 mph actually has membership to 2 fuzzy sets simultaneously (slow and average); however, the value of the membership function to both is less than 1. By classifying numerical input data using such linguistic terms ("fuzzyfying" data), control systems can be formulated by using if–then rule-based inference, e.g. if a car's speed is fast and an obstruction lies ahead then apply brakes hard. A typical control system is constructed of tens of such rules, maybe even more. At each iteration input variables are "fuzzified", each rule is then evaluated, and the output control signals are calculated as an amalgamation of all rule outputs. Many different ways exist to amalgamate rule outputs, and the method chosen is the own preference of a designer.

Fuzzy logic provides a remarkably simple way to draw definite conclusions from vague, ambiguous or imprecise information. In a sense, FL resembles human decision making with its ability to work with approximate data and find precise solutions. FL can be used in several different ways to implement a diagnosis system, and the most common one is to construct a set of rules from which conclusions and actions can be drawn. Several examples of how this technique can be applied to transformer DGA can be found in the following papers: Tomsovic et al. used fuzzy rules to represent existing DGA methods and then combined the results to form an overall diagnosis [12]; whereas, Yang [13] and Huang [14] employed FL to develop new rules for analysing DGA data, which were then evaluated on test sets and properties of memberships functions they use, tuned to improve performance.

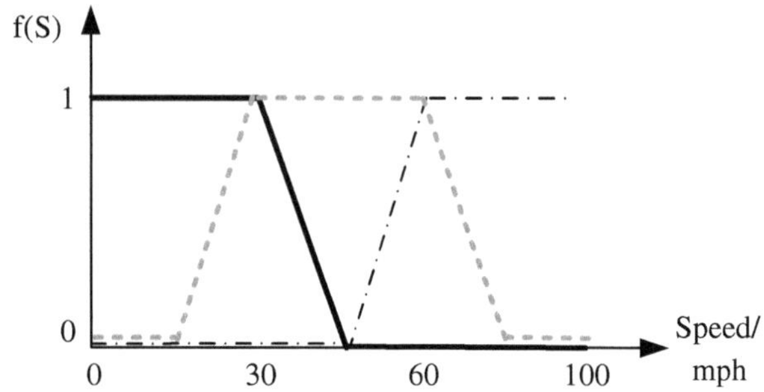

Fig. 3.4 Fuzzy sets describing a car's speed as slow, average and fast

3.4 Bayesian Networks

Conventionally, subjective data are handled by traditional probability theories like the Bayes' theorem. One drawback of such methods is that they require the values of a large number of conditional probabilities, which are often practically very difficult to obtain. Over the past decade, BNs [15, 16] have emerged as an increasingly active area for research in CI, which have covered a wide range of applications [17], e.g. medical diagnosis, expert systems, planning and learning. As a well-studied statistical tool, a Bayes net is a network consisting of a set of nodes and links between each pair of nodes, which can model joint probability distributions between the nodes based upon their conditional independence. As an enhancement to overcome the mentioned drawbacks, BNs provide a tool for probability learning from data and probabilistic inference, which are a marriage between the probability theory and the graph theory.

3.4.1 The Bayes' Theorem

The Bayes' theorem is the foundation of BN reasoning, which is a simple mathematical formula used for calculating conditional probabilities. It is named after Rev. Thomas Bayes, an eighteenth century mathematician who derived a special case of this theorem [18]. This leads to a common form of the Bayes' theory, Eq. 3.39, which allows us to compute the probability of one event in terms of observations of another and knowledge of joint distributions.

$$P(h|e) = \frac{P(e|h) \times P(h)}{P(e)}, \tag{3.39}$$

where $P(h|e)$ denotes the posterior probability of a hypothesis h conditioned upon some evidence e, $P(h)$ the prior probability of h, $P(e|h)$ the likelihood for e given h and $P(e)$ the prior or marginal probability of e. Therefore, the Bayes' theorem can be clearly interpreted as an alternative form in Eq. 3.40 with respect to each item in Eq. 3.39:

$$\text{Posterior Probability} = \frac{\text{Likelihood} \times \text{Prior Probability}}{\text{Probability of Evidence}}. \quad (3.40)$$

The Bayes' theorem asserts that the probability of a hypothesis h conditioned upon some evidence e, i.e. $P(h|e)$, is equal to its likelihood $P(e|h)$ times its probability prior to any evidence $P(h)$, and divided by $P(e)$ (so that the conditional probabilities of all hypotheses are summed to 1). In the BN theory, the Bayes' theorem is utilised to update the probabilities of variables whose state has not been observed given a set of new observations. Hence, the Bayes' rule allows unknown probabilities to be calculated from known cases.

3.4.2 Bayesian Networks

A BN is a graphical structure that allows us to represent and reason about an uncertain domain [17]. A typical BN is depicted in Fig. 3.5, which comprises a set of nodes and links between each pair nodes. The nodes in a BN represent a set of random variables in a particular problem domain. A set of directed arcs (or links) connects pairs of nodes, representing the direct dependencies between variables. Considering discrete variables in most cases, the strength of the relationship between variables is quantified by conditional probability distributions (CPD) associated with each node.

Most commonly, a BN is considered to be a representation of joint probability distributions. It is assumed that there is a useful underlying structure to the problem being modelled that can be captured by a BN. If such a domain structure exists, a BN gives a more compact representation than simply describing the probability of every joint instantiation of all variables. Consider a BN with n nodes as shown in Fig. 3.5, and X_1 to X_n are taken in such an order from 1 to 6.

A particular value in the joint distribution is represented by $P(X_1 = x_1, X_2 = x_2, \ldots, X_n = x_n)$, or more compactly, $P(x_1, x_2, \ldots, x_n)$. The chain rule of probability theory allows us to factorise joint probabilities, therefore:

Fig. 3.5 An example graphic model of Bayesian networks

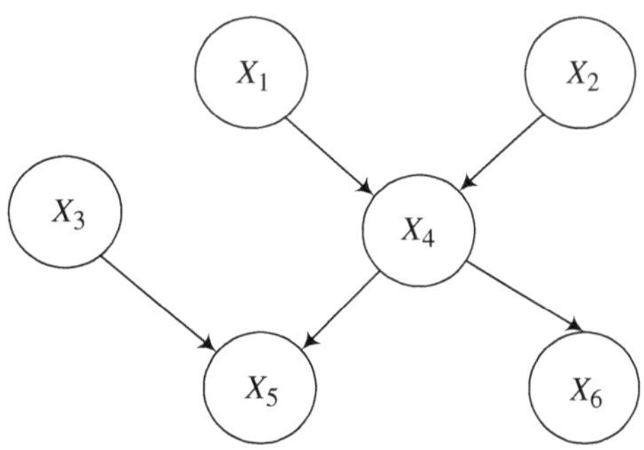

$$
\begin{aligned}
P(x_1, x_2, \ldots, x_n) &= P(x_1) \times \cdots \times P(x_n | x_1, \ldots, x_{n-1}) \\
&= \prod_{i=1}^{n} P(x_i | x_1, \ldots, x_{n-1}).
\end{aligned}
\tag{3.41}
$$

The structure of a BN implies that the value of a particular node is conditional only on the values of its parent nodes, then Eq. 3.41 is reduced to:

$$
P(x_1, x_2, \ldots, x_n) = \prod_{i=1}^{n} P(x_i | \mathrm{Parents(X_i)}). \tag{3.42}
$$

Once the topology of a BN is specified, the conditional probability table (CPT) for each node should be defined. It is defined that each row of a CPT contains the conditional probability of each node for each possible combination of values of its parent nodes. Each row of a CPT must sum to 1, and the values of 0 and 1 mean that the probability of one hypothesis upon its evidence is 0 or 1. It is noted that, a node with no parents has only prior probabilities. Then, probabilistic inference with BNs is achieved by updating beliefs—that is, computing the posterior probability distributions—given new evidence. The essential idea is that new evidence has to be propagated to other parts of a BN.

3.4.3 Parameter Learning to Form a Bayesian Network

In order to describe a BN and perform probabilistic inference, the graph structure and the parameters of each CPT should be specified. As in most cases, structures of BNs can be determined with help from expert assistance. However, the prior conditional probabilities in each CPT are difficult to obtain in real-world applications. An obvious solution is to employ automated machine learning to discover parameters of each CPT, which is beyond the main scope of this book. Therefore, only essentials of a parameter learning process for BNs are presented [17] as below.

Learning parameters from historical data is always regarded as very important. In the applications with known structure and full observability, the objective of learning is to find the parameters of each CPT, which maximises the likelihood of training data. For example, the training data contain N cases, which are assumed to be independent. The normalised log-likelihood L of the training set D is a sum of terms for each node:

$$
\max L = \frac{1}{N} \sum_{i=1}^{n} \sum_{D} \log P(X_i | \mathrm{Parent(X_i), D}). \tag{3.43}
$$

It is noted that the log-likelihood scoring function decomposes according to the structure of a graph, and hence, we can maximise the contribution to the log-likelihood of each node independently. When the model structure and CPT parameters are specified, probabilistic inference amongst all variables can be accomplished using the created BN. More discussions can be found in [17] addressing parameter and structure learning for various types of BNs.

3.5 Summary

This chapter focusses on the three mathematical theories dealing with uncertainties, i.e. the ER theory, the FL theory and BNs. First, an ER algorithm, which can be implemented to deal with uncertain decision knowledge for MADM problems, is described. The ER approach is regarded as a decision-making method established to provide an overall decision by aggregating subjective judgements on constitutive attributes of an MADM problem involving uncertainties. Then, the principle of the FL theory is discussed alongside with a simple example. The membership functions in the FL theory are applied to "soften" crisp fault decision boundaries in Chap. 8. Finally, the foundation of BNs is illustrated, which are represented as directed acyclic graphs. The nodes and links of the graphs represent variables and encode conditional independencies between variables, respectively. Generalisations of BNs, which can represent and solve decision problems involving uncertainties, are deployed as a classifier for diagnosing transformer faults in Chap. 8.

References

1. Finlay J, Dix A (1996) An introduction to artificial intelligence, 3rd edn. UCL Press, London
2. Zadeh LA (1965) Fuzzy sets. Inf Control 8:338–353
3. Negnevitsky M (2005) Artificial intelligence: a guide to intelligent systems, 2nd edn. Pearson Education, England
4. Ghosh JK, Delampady M, Samanta T (2006) An introduction to bayesian analysis: theory and methods. Springer, New York
5. Shafer G (1976) A mathematical theory of evidence. Princeton University Press, Princeton
6. Yang JB, Singh MG (1994) An evidential reasoning approach for multiple attribute decision making with uncertainty. IEEE Trans Syst Man Cybern 24(1):1–18
7. Yang JB, Sen P (1994) A general multi-level evaluation process for hybrid MADM with uncertainty. IEEE Trans Syst Man Cybern 24(10):1458–1473
8. Tang WH, Wu QH, Richardson ZJ (2004) An evidential reasoning approach to transformer condition monitoring. IEEE Trans Power Deliv 19(4):1696–1703
9. Spurgeon K, Tang WH, Wu QH, Richardson ZJ, Moss G (2005) Dissolved gas analysis using evidential reasoning. IEE Proc Sci Meas Technol 152(3):110–117
10. Yang JB, Xu DL (2002) On the evidential reasoning algorithm for multiple attribute decision making under uncertainty. IEEE Trans Syst Man Cybern A Syst Hum 32(3):289–304
11. Buchanan BG, Shortliffe EH (1984) Rule-based expert systems. Addison-Welsey, Reading
12. Tomsovic K, Tapper M, Ingvarsson T (1993) A fuzzy approach to integrating different transformer diagnostic methods. IEEE Trans Power Deliv 8(3):1638–1646
13. Yang HT, Liao CC (1999) Adaptive fuzzy diagnosis system for dissolved gas analysis of power transformers. IEEE Trans Power Deliv 14:1342–1350

14. Huang YC, Yang HZ, Huang CL (1997) Developing a new transformer fault diagnosis system through evolutionary fuzzy logic. IEEE Trans Power Deliv 12(2):761–767
15. Friedman N, Linial M, Nachman I, Pe'er D (2000) Using Bayesian networks to analyze expression data. J Comput Biol 7:601–620
16. Pearl J (1988) Probabilistic reasoning in intelligent systems. Morgan Kaufmann Press, San Francisco
17. Korb BK, Nicholson AE (2003) Bayesian artificial intelligence. Chapman & Hall/CRC Press, London
18. Bayes RT (1958) An essay toward solving a problem in the doctrine of chances. Philos Trans R Soc Lond 53:370–418 (1763), reprinted in Biometrika 45:293–315 (1958)

Chapter 4
Thermoelectric Analogy Thermal Models of Power Transformers

Abstract This chapter first presents the conventional thermal models adopted in the IEC and the IEEE regulations for oil-immersed power transformers, including steady-state models, transient-state models and hot-spot temperature rise models. In order to improve the calculation accuracy of transformer thermal models, two thermoelectric analogy thermal models have been developed, which are rooted on the principles of heat exchange and electric circuit laws, including a full thermoelectric analogy model and a simplified thermoelectric analogy model. Two sets of differential equations are derived to calculate both the transient-state temperatures and the stationary-state equilibria of the main parts of a transformer. Finally, a practical means is derived to calculate hot-spot temperatures based on the outputs of the two proposed thermoelectric analogy models.

4.1 Introduction

The useful life of a transformer is determined partially by its ability to dissipate the internally generated heat to its surroundings. All oil-immersed power transformers are designed to meet certain operating criteria with regard to temperatures. The thermal capacity of oil and windings permits operations above-rated loads for short periods. This is usually what is meant by thermal ratings, for instance:

1. Mean winding temperature rise (above ambient) <65°C at a rated load.
2. Top oil temperature rise <60°C at a rated load.
3. Hot-spot temperatures cannot exceed 125°C at ambient temperature of 20°C.
4. Cyclic capability of 1.2 per unit.

Improved knowledge about thermal characteristics of transformers enables enhanced transformer ratings and reduces risks associated with emergency post-fault operations. The development of accurate transformer thermal models is

W. H. Tang and Q. H. Wu, *Condition Monitoring and Assessment of Power Transformers Using Computational Intelligence*, Power Systems,
DOI: 10.1007/978-0-85729-052-6_4,

always regarded as one of the most important issues of transformer condition monitoring [1, 2]. A comparison of real and predicted operation temperatures can provide a sensitive diagnosis of transformer conditions, which may indicate a fault. Due to the drawbacks of the traditional thermal solutions stated in Sect. 1.3.1, many attempts have been made to improve the calculation accuracy of transformer thermal models.

In [3], a radial basis function neural network (RBFN) was designed and trained to estimate winding hot-spot temperatures, taking into account the influence of weather on thermal behaviours of a transformer. The RBFN model provided a good mapping between model inputs and outputs. However, it does not possess any physical meaning, noted for direct mapping, and the model can only represent the relationship between model inputs and outputs accurately within the range covered by training data. Hence, the extrapolability of the RBFN model is very limited [4]. Another approach based on exponential responses of oil temperatures [5], derived several differential equations, which were used to predict transformer temperatures complying to empirical equations. The thermal parameters in these equations were optimised with a nonlinear least square optimisation technique [5]. Its capability to predict transformer temperatures during realistic cooler switching conditions is limited while cooler states are on and off, as the nonlinearity of thermal dynamics could not be taken into account in the derived differential equations. To sum up, an accurate thermal model, which can reflect the nonlinearity of transformer thermal dynamics, is highly desirable.

4.2 Conventional Thermal Models in IEC and IEEE Regulations

4.2.1 Steady-State Temperature Models

For un-pumped (or natural) oil cooling of a power transformer (ON), the oil temperature at the top of a winding is approximately equal to the TOT inside its tank. However, for forced oil circulation (OF), the TOT is the sum of the oil temperature at the bottom of the winding, BOT, and the difference between the oil temperatures at the top and bottom of the winding [2].

The steady-state bottom-oil temperature is described as follows:

$$\theta'_{\mathrm{BO}} = \theta_{\mathrm{a}} + \theta_{\mathrm{bo}}\left(\frac{1 + dK^2}{1 + d}\right)^x, \tag{4.1}$$

and the steady-state top-oil temperature is represented by:

$$\theta'_{\mathrm{TO}} = \theta_{\mathrm{a}} + \theta_{\mathrm{bo}}\left(\frac{1 + dK^2}{1 + d}\right)^x + (\theta_{\mathrm{to}} - \theta_{\mathrm{bo}})K^y, \tag{4.2}$$

where θ_a is the ambient temperature; K the ratio of the operating load current to the rated load current, $K = \frac{I_C}{I_R}$; θ'_{BO} the steady-state oil temperature at the bottom of a winding with a load ratio K; θ_{bo} the oil temperature rise above ambient at the bottom of a winding ($K = 1$); d the ratio of load loss (at a rated load) to no-load loss; θ'_{TO} the steady-state oil temperature at the top of a winding with a load ratio K; θ_{to} the oil temperature rise above ambient at the top of a winding ($K = 1$); x the exponent related to oil temperature rise due to total losses; y the exponent related to winding temperature rise due to load currents.

The model parameters, e.g. d, x and y, are usually determined through experiments or by experience.

4.2.2 Transient-State Temperature Models

In contrast to winding temperatures, the transient-state BOT and TOT cannot immediately reach the corresponding steady-state states under changing loads, since the thermal time constants of BOT and TOT are of the order of hours.

Conventionally, any change in load conditions is treated as a step function. The rectangular load profile considered in the loading tables of IEC60354 consists of a single step up followed by a single step down after a predefined time interval [6]. For a continually varying load, a step function is applied over a small time interval. In this way, the transient-state BOT and TOT at any time instant t during a loading period i are given by the following two equations. For the transient-state BOT:

$$\theta'_{bo}(t) = \theta'_{bo(i-1)} + (\theta'_{boi} - \theta'_{bo(i-1)})(1 - e^{\frac{-t}{\tau_{bo}}}), \tag{4.3}$$

and for the transient-state TOT:

$$\theta'_{to}(t) = \theta'_{to(i-1)} + (\theta'_{toi} - \theta'_{to(i-1)})(1 - e^{\frac{-t}{\tau_{to}}}), \tag{4.4}$$

where τ_{bo} and τ_{to} are the thermal time constants of BOT and TOT, respectively, and $\theta'_{bo(i-1)}$ and θ'_{boi} the BOTs at the beginning and end of section i of a loading chart respectively derived from Eq. 4.1. Similar definitions are applied to $\theta'_{to(i-1)}$ and θ'_{toi}, which are derived from Eq. 4.2.

However, the above calculation of temperatures is dependent upon the estimation of thermal time constants, i.e. τ_{bo} and τ_{to}, which are normally obtained from historical experiments and not always accurate. Moreover, the traditional transient models can only calculate a loading profile with step functions, which cannot be applied to a loading scenario with continuous changes.

4.2.3 Hot-Spot Temperature Rise in Steady State

The most critical limitation, when operating a transformer, is the temperature reached in the hottest area of a winding, and every effort should be made to determine accurately this temperature. A direct measurement (with a fibre-optic probe or similar devices) is now becoming available. Such devices can directly measure HSTs, which can be compared with the HST calculations derived from a conventional thermal model for fault diagnosis purposes. However, most of the on-site transformers are not equipped with such a direct measurement probe, and HSTs are usually estimated with oversimplified thermal characteristics as presented below.

4.2.3.1 Assumed Thermal Characteristics with Simplifications

The maximum temperature occurring in any part of a winding insulation system is called the hot-spot temperature. Investigations have shown that the top-oil temperature inside a winding might be 5–15 K higher than that of the mixed top-oil inside a tank. The actual temperature difference between conductor and oil is assumed to be higher by a hot-spot factor.

A typical temperature distribution is assumed as shown in Fig. 4.1, on the understanding that such a diagram is the simplification of a more complex temperature distribution. The assumptions made in this simplification are as follows [6]:

1. The oil temperature inside a winding increases linearly from bottom to top, whatever the cooling mode be.
2. The temperature rise of the conductor at any vertical position of a winding increases linearly, parallel to the oil temperature rise, with a constant difference g between the two straight lines (g being the difference between the average temperature rise by resistance and the average oil temperature rise).
3. The hot-spot temperature rise is higher than the temperature rise of the conductor at the top of a winding as shown in Fig. 4.1, because allowance has to be made for the increase in stray losses. To take account of these nonlinearities,

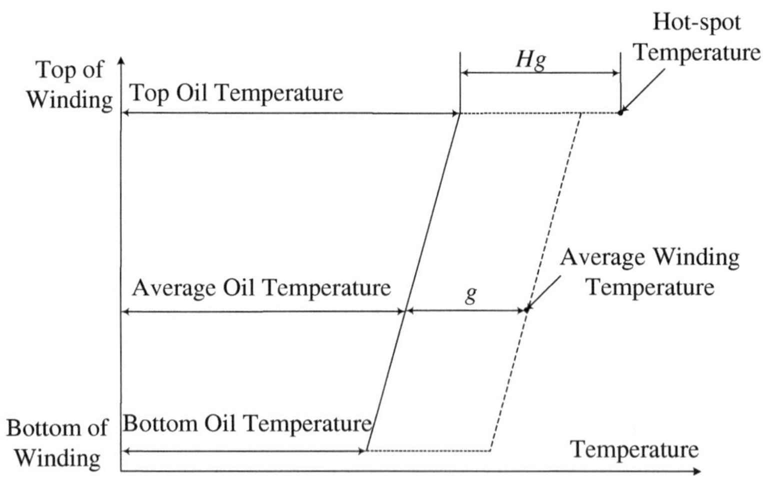

Fig. 4.1 Typical temperature distribution of a transformer

the difference in temperatures between the hot-spot and the oil at the top of the winding is made equal to *Hg*. The hot-spot factor *H* may vary from 1.1 to 1.5 depending on the following factors, e.g. transformer size, short-circuit impedance and winding design.

In order to calculate the HST rise under continuous, cyclic or other duties, different sources of thermal characteristics are employed:

1. Results of a special temperature-rise test including the direct measurement of HST or TOT inside a winding (in the absence of direct hot-spot measurement, *H* can only be provided by a manufacturer).
2. Results of a normal temperature rise test.
3. Assumed temperature rises at the rated current.

4.2.3.2 ON Cooling

For ON cooling, the ultimate hot-spot temperature, θ_h, under any load *K* is equal to the sum of the ambient temperature, the top-oil temperature rise and the temperature difference between the hot-spot and the top-oil:

$$\theta_h = \theta_a + \Delta\theta_{or}\left(\frac{1+RK^2}{1+R}\right)^x + Hg_rK^y, \tag{4.5}$$

where $\Delta\theta_{or}$ is the top-of-winding oil temperature rise when $K = 1$, *R* the loss ratio and Hg_r the hot-spot to top-oil gradient when $K = 1$.

4.2.3.3 OF Cooling

For OF cooling, the calculation method is based on the bottom-oil and average oil temperatures. Thus the ultimate hot-spot temperature, θ_h, under any load *K* is equal to the sum of the ambient temperature, the bottom-oil temperature rise, the difference between the top-oil in the winding and the bottom-oil and the difference between the hot-spot and the top-oil in a winding:

$$\theta_h = \theta_a + \Delta\theta_{br}\left(\frac{1+RK^2}{1+R}\right)^x + 2[\Delta\theta_{imr} - \Delta\theta_{br}]K^y + Hg_rK^y, \tag{4.6}$$

where $\Delta\theta_{br}$ is the bottom-oil temperature rise when $K = 1$ and $\Delta\theta_{imr}$ the average oil temperature rise when $K = 1$.

4.2.3.4 OD Cooling

For oil-directed (OD) cooling, the calculation method is basically the same as for OF cooling except that a correction term is added to take into account variations of the ohmic resistance of conductors with temperatures:

$$\theta_{\mathrm{h}}' = \theta_{\mathrm{h}} + 0.15(\theta_{\mathrm{h}} - \theta_{\mathrm{hr}}) \quad (K > 1), \tag{4.7}$$

where θ_{h}' is the ultimate hot-spot temperature with a correction term, θ_{h} the one without consideration of the influence of the ohmic resistance variations using Eq. 4.6 and θ_{hr} the hot-spot temperature at rated conditions.

Table 4.1 provides typical values of thermal characteristics reported by IEC60354 for oil-immersed power transformers.

4.2.4 Hot-Spot Temperature Rise in Transient State

Due to that transformer loading ratios change from time to time, the winding HST rise over TOT can jump immediately between the corresponding steady-state values. It is considered that the thermal time constant of HST is very small, just of the order of minutes (typically between 5 and 6 min). As recommended by IEC, there are two ways to calculate HSTs as a function of time for varying load currents and ambient temperatures [2]:

1. The exponential equation solution, suitable for a load variation according to a step function, is particularly suited for the determination of heat transfer parameters through tests, especially by manufacturers.
2. The difference equation solution, suitable for arbitrarily time-varying load current K and time-varying ambient temperature θ_{a}, is particularly applicable for on-line monitoring. Therefore, this method is adopted in Sect. 4.7.2 to calculate HSTs for on-line condition monitoring purposes.

4.3 The Thermoelectric Analogy Theory

Heat is a form of invisible energy which is always transferred between two communicating systems, arising solely from a temperature difference. The rate of

Table 4.1 Thermal characteristic listed in IEC60354

	Distribution transformer	Medium and large size unit		
	ONAN	ON	OF	OD
x	0.8	0.9	1.0	1.0
y	1.6	1.6	1.6	2.0
R	5.0	6.0	6.0	6.0
H	1.1	1.3	1.3	1.3
τ_{o} (h)	3.0	2.5	1.5	1.5
θ_{a} (°C)	20	20	20	20
$\Delta\theta_{\mathrm{hr}}$ (K)	78	78	78	78
Hg_{r} (K)	23	26	22	29
$\Delta\theta_{\mathrm{imr}}$ (K)	44	43	46	46
$\Delta\theta_{\mathrm{br}}$ (K)	33	34	36	43

heat flow depends mainly upon certain physical properties of an observed body, e.g. respective temperatures and the magnitude of their difference, as well as ambient conditions. In simple cases, it can be determined quantitatively by applying the basic principles of thermodynamics and fluid mechanics. The Fourier theory, which is one of the most well-known fundamental laws of heat transfer, is written as the following [7]:

$$Q = \frac{kS\Delta T}{\delta} = \frac{\Delta T}{R_{\mathrm{h}}}, \tag{4.8}$$

where Q denotes the heat flow (W), ΔT the temperature difference (K), R_{h} the heat resistance (K/W), S the area in contact with heat (m^2), k the thermal conductivity (W/mK) and δ the length of heat flow path (m). The heat transfer equation has the same form as the Ohm's law used for electrical circuits, which is described as:

$$I = \frac{E}{R_{\mathrm{e}}}, \tag{4.9}$$

where I is the electrical current (A), E the electrical potential difference (V) and R_{e} the electrical resistance (Ω).

With a comparison of the above two fundamental laws, the basic equations for heat transfer and electrical charge transport are analogous. As a result, in many cases, the mathematical description of a heat transfer system is similar to that of an electrical system. Due to the mathematical similarity between heat transfer and electric charge transport, an equivalent electrical network model can be used to solve a complicated heat transfer problem. It can not only handle the nonlinearity and change of heat transfer network effects, but also offers a useful calculation method that has been rather neglected by researchers and other specialists [8]. The analogy between variables of the thermal and electrical fields is listed in Table 4.2, and that between the thermal and the electrical constants in Table 4.3, where c_p is the specific heat capacity and ρ and v are the density and volume of a studied component, respectively.

4.4 A Comprehensive Thermoelectric Analogy Thermal Model

4.4.1 Heat Transfer Schematics of Transformers

An oil-immersed power transformer normally consists of a pair of windings, primary and secondary, linked by a magnetic circuit or a core, which are immersed in transformer oil that both cools and insulates windings. When resistive and other losses are generated in a transformer winding, heat is produced. This heat is then transferred into and taken away by the insulation oil. Although winding copper materials retain mechanical strength up to several hundred degrees Celsius,

Table 4.2 Analogy between variables of thermal and electrical fields

Thermal field	Electrical field
H, quantity of heat	*M*, quantity of electricity
Q, rate of heat flow	*I*, electrical current
T, temperature rise	*E*, electrical potential difference

Table 4.3 Analogy between constants of thermal and electrical fields

Thermal field
kS, product of thermal conductivity and area
$c_p\rho v$, thermal capacity
$k/c_p\rho$, thermal diffusivity
Electrical field
$1/R_e$, electrical conductance
C_e, electrical capacitance
$1/R_eC_e$, ratio of conductance to capacitance

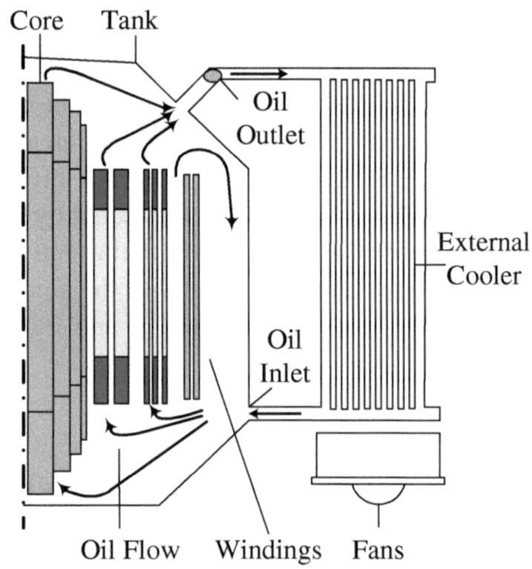

Fig. 4.2 Schematics of active parts of an oil-immersed transformer

transformer oil degrades significantly with temperatures above 140°C and paper insulation also deteriorates with greatly increasing severity if its temperature rises are above 90°C. The cooling oil flow must, therefore, ensure that the temperature of insulation materials is kept below this figure as far as possible. The schematics of main parts of an oil-immersed transformer are shown in Fig. 4.2 comprising a core, windings, the insulation oil, an oil inlet, an oil outlet, a tank and external coolers.

To investigate transformer thermal dynamics, a transformer is expressed in an abstractive manner in Fig. 4.3 referring to Fig. 4.2, in which the main parts of a

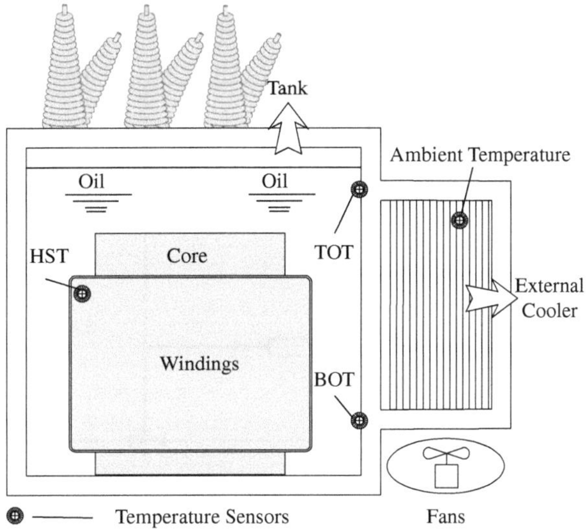

Fig. 4.3 Abstractive schematics of a transformer and nomenclature

transformer and their nomenclature are illustrated. It can be noticed that four types of temperatures are listed, i.e. the ambient temperature, BOT, TOT and HST. With respect to Fig. 4.3, the essential heat transfer schematics of a transformer are explained below.

The heat generated by resistive and other losses in a power transformer is transferred into and taken away by transformer oil through the surfaces of an oil tank and a cooler in three ways—convection, conduction and radiation. Part of the heat exchange schematics of an oil natural air natural and oil force air force (ONAN/OFAF) power transformer is illustrated in Fig. 4.4, where Q_{Cu} denotes the heat generated by copper losses, Q_{Fe} the heat generated by core losses, T_{hs} the hot-spot temperature, T_s the tank and cooler surface temperature, T_e the environment temperature, G''_{12} the heat conductance due to the heat exchange between the winding and the core, G''_1 the heat conductance of the winding to the oil, G''_2 the heat conductance of the core to the oil and G''_3 the heat conductance of the oil to the tank and the cooling medium.

4.4.2 Derivation of a Comprehensive Heat Equivalent Circuit

Due to the analogy between electric charge transport and one-dimensional heat flow [7, 8], it is convenient to set up an equivalent heat circuit corresponding to actual thermo-hydraulic structures, and to adjust thermal parameters for actual areas, which is detailed in [9]. Generally, an equivalent heat circuit consists of heat conductors, heat capacitors and heat current sources. In establishing such an equivalent heat circuit for an ONAN/OFAF power transformer, a thermoelectric analogy thermal model is developed by regarding the active heat transfer parts of a

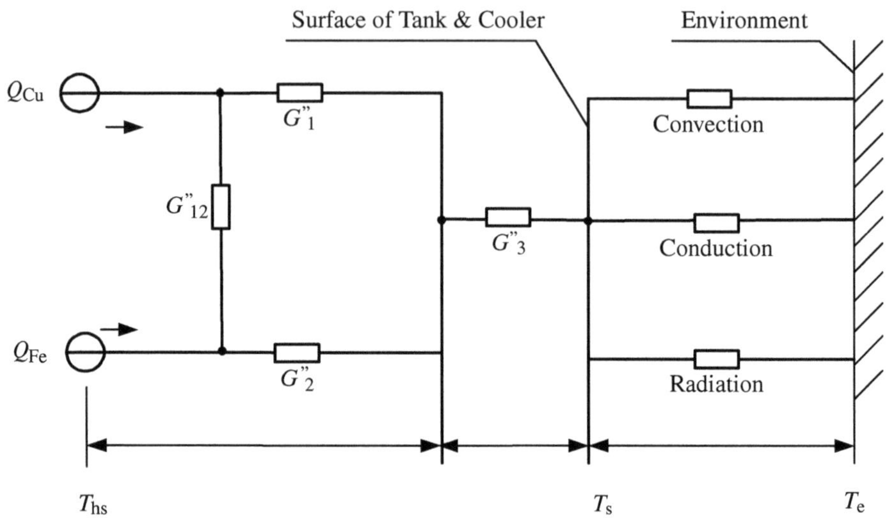

Fig. 4.4 Part of the transformer heat exchange schematics

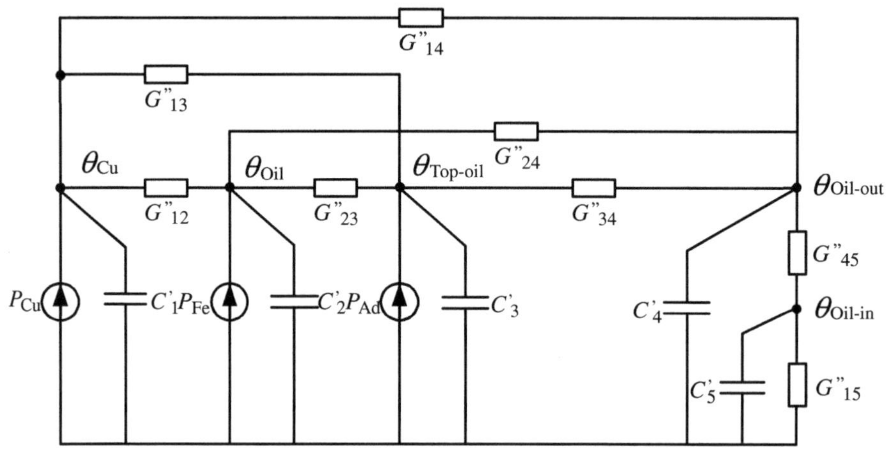

Fig. 4.5 A full equivalent heat circuit of an oil-immersed power transformer

transformer as four lumped components, including a coil assembly, the insulation oil, a transformer tank and an external cooler.

For a global heat flow description, individual parts are denoted by corresponding nodes, representing the complete volume of a part of a transformer. The paths of heat convection, diffusion and radiation are simulated by heat resistances to the flow of heat. As a result, distributed heat sources and heat resistances are represented by several lumped heat sources, equivalent thermal resistors and overall thermal capacitors [7, 10]. Then, a full heat equivalent circuit of an oil-immersed power transformer is developed as illustrated in Fig. 4.5, where P_{Cu}

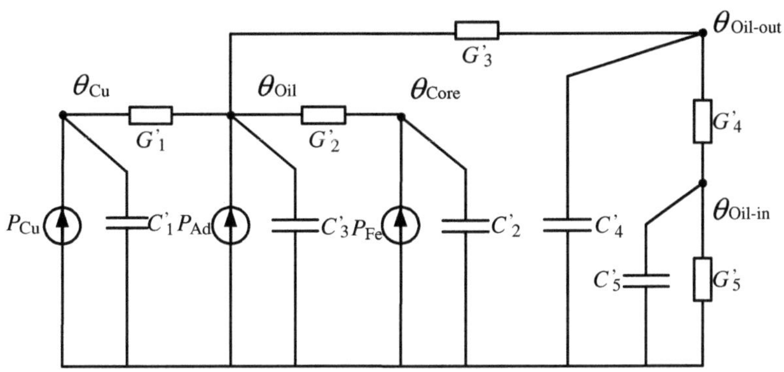

Fig. 4.6 A comprehensive equivalent heat circuit of an oil-immersed power transformer

denotes winding copper losses, P_{Fe} core losses, P_{Ad} stray losses, G_{ij}'' the thermal conductance between a pair of nodes, C_i' the thermal capacitance of each node and θ the temperature rise of each node.

However the circuit in Fig. 4.5 is too complicated, as the effects of some components are negligible in a practical sense. After simplification using network reduction techniques, a comprehensive thermoelectric analogy thermal model (CTEATM) is shown in Fig. 4.6, where $\theta_{\text{Oil-out}}$ is the temperature rise at the oil outlet, $\theta_{\text{Oil-in}}$ the temperature rise at the oil inlet, G_i' the concentrated thermal conductance, C_i' the overall thermal capacitance of each node, θ_{Cu} the average temperature rise of the winding, θ_{Core} the average temperature rise of the core, θ_{Oil} the average temperature rise of the oil.

In Fig. 4.6, the circuit operates subject to the Kirchhoff's law: currents into a node sum to zero, so do voltages around a mesh. Therefore, the analysis of such a network leads to the solution of a matrix of n equations with n unknowns, where n is the number of nodes. These nodal equations correspond exactly to the finite difference equations derived from the thermodynamic differential equations. As a result of subdividing a transformer into four lumped components, the calculated nodal temperatures in Fig. 4.6 are the average temperatures of each main part. CTEATM can then be employed to predict BOTs and TOTs using a set of ordinary, first-order differential equations. After comparing the matrix of thermodynamics equations with the basic electrical principles, the following differential equations are derived for calculating BOTs and TOTs.

$$\mathbf{P}\mathrm{d}t = \mathbf{G}'\Theta\mathrm{d}t + c'md\Theta, \tag{4.10}$$

or

$$\mathbf{P} = \mathbf{G}'\Theta + \mathbf{C}'\mathrm{d}\Theta/\mathrm{d}t, \tag{4.11}$$

where $\mathbf{P}$ is the heat power input vector, $\mathbf{G}'$ the heat conductance matrix, $\mathbf{C}'$ the heat capacitance matrix, Θ the temperature rise vector (above ambient temperatures), c' the specific heat capacity vector and m the object mass vector. In Eq. 4.11, $\mathbf{P}$ is the time-dependent energy source and once a set of $\mathbf{G}'$ and $\mathbf{C}'$ has been determined, the major unknown is Θ, which can then be derived by solving the differential equation (4.11) with the Runge–Kutta method.

4.5 Parameter Estimation of a Thermoelectric Analogy Model

4.5.1 Heat Generation Process

Heat is generated in the core and coil assembly in the form of three types of losses: the copper losses, which result from I^2R_e, the stray losses in the winding and the iron losses, which are the sum of hysteresis and eddy-current power. In order to get the heat power input of the developed thermoelectric analogy model, copper losses and core losses of a power transformer are to be derived with varying load currents and temperatures. Generally, the current flow in any electrical system is dependent upon the magnitude of the current source and the resistance of the system. Transformer windings are no exception and these give rise to the copper losses of a transformer. Copper losses are proportional to the square of a load current [2] as:

$$P_{\mathrm{Cu,C}} = P_{\mathrm{Cu,N}} \left(\frac{I_{\mathrm{C}}}{I_{\mathrm{N}}}\right)^2 \left(\frac{235 + \theta_{\mathrm{C}}}{235 + \theta_{\mathrm{N}}}\right), \tag{4.12}$$

where $P_{\mathrm{Cu,C}}$ represents the resistive loss at an operating load current, $P_{\mathrm{Cu,N}}$ the resistive loss at a rated load, I_{C} the operating load current, I_{N} the rated load current, θ_{C} the current winding temperature and θ_{N} the winding temperature at the rated losses. The magnetising current is required to take a core through the alternating cycles of flux at a rate determined by system frequencies. This is known as the core loss. In this study, the core loss is considered as a constant obtained from a factory test. The stray losses are caused by the stray fluxes in windings and core clamps, which are readily available from a transformer handbook as a proportion of the total load loss.

4.5.2 Heat Transfer Parameter

Based upon the density, volume and the specific heat of an observed part of a transformer, the approximate range of thermal capacitance per unit length is calculated with the following equation

$$C = c_p \rho v. \tag{4.13}$$

Each thermal capacitance is only related to one temperature node, as there is no coupling capacitance between the nodes in a heat equivalent circuit.

Using Eq. 4.13, the thermal capacitance range of each lumped component can be estimated. As for the nonlinear thermal conductances, they are related to various factors of a power transformer, such as geometrical size, temperature and heat characteristics. It is therefore very important to choose a correct form to represent their nonlinear nature. In many cases, an exponential form is adopted which possesses a simple format yet with a satisfactory accuracy:

$$G = a\Delta T^b. \tag{4.14}$$

In Eq. 4.14, ΔT denotes the temperature difference or temperature rise, and a and b are coefficients to be determined for a heat transfer path.

4.5.3 Operation Scheme of Winding Temperature Indicator

Most of large power transformers in transmission networks are designed to be operated in certain circumstances at loadings in excess of their rating. Because individual transformers vary, a winding temperature indicator (WTI) is normally used for loading transformers under different cooling schemes. In practice, forced cooling of transformers is controlled automatically according to winding temperatures measured by a fitted WTI. When the winding temperature increases and exceeds a predefined value, θ_0', the pumps and fans are switched on to dissipate the generated heat to its sounding.

The general principle of setting a WTI is described briefly in this paragraph. In practice, the hottest spot temperature of a winding is simulated by increasing the indication on the thermometer measuring the top oil temperature by injecting current into a heater fitted to the instrument operating bellows. The heater current, and hence the instrument reading, is controlled by varying the value of a resistor shunting the heater coil. A WTI must be set to read the transformer top oil temperature adding the product of the maximum winding gradient and the hot-spot factor. For example, suppose that the secondary current of a current transformer supplying the WTI heater circuit is 2.0 A at a rated current. The maximum winding difference is 34.6°C and H is 1.1, so that the heater in the WTI could boost the oil temperature by 38.1°C. The fans and pumps fitted to the cooling system of this transformer are arranged so that the motors can be switched on automatically in two groups under the control of WTI, or each motor can be switched on individually by hand. With the master control switch set to be automatic, the temperature indicator contacts will start up the motors as soon as a temperature of 75°C is exceeded. The motor will be switched off again when the temperature falls below 50°C.

4.5.4 Time Constant Variation in a Heat Transfer Process

From field experiences, the oil time constants are quite different in two cooling conditions, e.g. ONAN or OFAF, which are controller by a WTI. To distinguish the different thermal dynamics appropriate to the periods when a cooler is on and when it is off, a two-piece model is necessary. A qualitative explanation could be given on the basis of the fact that the volume distributions of temperature in the two processes are different, e.g. the active mass of the oil involved in the heat transfer is substantially different when pumps and fans are on and off, which causes variations of thermal parameters of CTEATM. However, more research is required to further study this thermal physics in the future.

Consequently, two sets of heat exchange parameters are required for conducting simulations of CTEATM, which can reflect changes of thermal time constants due to different loading and cooler operation conditions of an ONAN/OFAF power transformer. It is crucial to estimate the heat conductance **G'** and the heat capacitance **C'** involved in different cooler operation conditions. As the overall heat capacitance can be estimated using Eq. 4.13, the upper boundary of **C'** can be determined. Regarding the heat conductance **G'**, as the temperature equilibria under the two cooler operation conditions can be obtained from an off-line heat run test, using Eq. 4.14, **G'** can also be estimated under the two cooling states.

4.6 Identification of Thermal Model Parameters

After the derivation of the comprehensive thermoelectric thermal model, how to determine each parameter (i.e. **C'** and **G'**) of the model becomes the main problem. Normally, to get the values of thermodynamics parameters, a set of off-line experiments should be conducted. However, it is not practical to shut down a large on-line power transformer and do a heat run test. Additionally, such an approach does not provide satisfactory results, as general thermal parameters cannot be determined precisely enough due to the complexity of transformer thermal phenomena. In the next chapter, an evolutionary computation algorithm, i.e. GA [4, 11, 12], is employed to identify the model parameters based upon real-world measurements sampled from on-line power transformers.

4.7 A Simplified Thermoelectric Analogy Thermal Model

4.7.1 Derivation of a Simplified Heat Equivalent Circuit

A 5-node equivalent heat circuit for an oil-immersed power transformer has been fully discussed in Sect. 4.4.2 [9]. However, in the previously constructed 5-node model, a relatively large number of thermal parameters need to be optimised as a result of so many temperature nodes. Also, due to the limited on-line measurements available, not all the temperature nodes are involved as nodal inputs for the thermal parameter identification. Only the TOT and BOT are provided to derive the whole set of the thermal parameters, which might cause uncertainty and

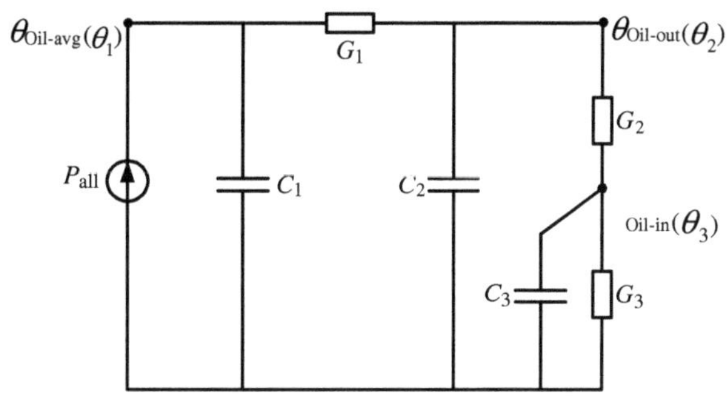

Fig. 4.7 A further simplified equivalent heat circuit of an oil-immersed power transformer

instability of a parameter optimisation process, and therefore a further simplification of the 5-node thermal model is necessary and practical. It is also considered that the model learning and predicting processes can be speeded up if the number of thermal parameters is reduced.

A further simplified thermoelectric analogy thermal model (STEATM) is shown in Fig. 4.7 [12]. First, the capacitors of core, windings and part of the surrounding oil are combined into one lumped capacitor C_1. Subsequently, the copper losses P_{Cu}, iron losses P_{Fe} and stray losses P_{Ad} are combined into a single heat source P_{all}. Accordingly the thermal conductances at relevant temperature nodes are united into one equivalent thermal conductance G_1, which is also in a nonlinear form. Then, leave the temperature nodes of TOT and BOT intact, as the temperature data for the two nodes can be measured conveniently, which provide a full description of the thermal behaviours of the two temperature nodes, while C_2, C_3, G_2 and G_3 are the relevant heat capacitances and conductances of the two temperature nodes, respectively.

The main parameters of STEATM are listed as below.

P_{all} overall losses, including P_{Cu}, P_{Fe} and P_{Ad};
G_1 overall winding and core to oil thermal conductance;
G_2 thermal conductance of coolers from the outlet to the inlet;
G_3 coolers to environment thermal conductance;
C_i concentrated thermal capacitance of each part of G_i;
θ_1 average oil temperature rise $\theta_{\mathrm{Oil\text{-}avg}}$;
θ_2 top-oil temperature rise $\theta_{\mathrm{Oil\text{-}out}}$ at the outlet;
θ_3 bottom-oil temperature rise $\theta_{\mathrm{Oil\text{-}in}}$ at the inlet.

As a result of the further simplification performed upon a transformer illustrated in Fig. 4.7, the calculated nodal temperatures are the average temperature rise of each main part, i.e. $\theta_{\mathrm{Oil\text{-}avg}}$, $\theta_{\mathrm{Oil\text{-}out}}$ and $\theta_{\mathrm{Oil\text{-}in}}$. STEATM can be employed to calculate $\theta_{\mathrm{Oil\text{-}in}}$ and $\theta_{\mathrm{Oil\text{-}out}}$, which are expressed by a set of ordinary, first-order and nonhomogenous differential equations. By comparing the matrix of thermodynamic equations with the basic electrical circuit principles, the computation is undertaken using the following differential equations:

$$\mathbf{P} = \mathbf{C}\mathrm{d}\Theta/\mathrm{d}t + \mathbf{G}\Theta, \tag{4.15}$$

$$\begin{bmatrix} P_{\text{all}} \\ 0 \\ 0 \end{bmatrix} = \begin{bmatrix} C_1 & 0 & 0 \\ 0 & C_2 & 0 \\ 0 & 0 & C_3 \end{bmatrix} \begin{bmatrix} \mathrm{d}\theta_1/\mathrm{d}t \\ \mathrm{d}\theta_2/\mathrm{d}t \\ \mathrm{d}\theta_3/\mathrm{d}t \end{bmatrix} + \begin{bmatrix} G_1 & -G_1 & 0 \\ -G_1 & G_1 + G_2 & -G_2 \\ 0 & -G_2 & G_2 + G_3 \end{bmatrix} \begin{bmatrix} \theta_1 \\ \theta_2 \\ \theta_3 \end{bmatrix}. \tag{4.16}$$

4.7.2 Hot-Spot Temperature Calculation

Hot-spot temperature is the maximum temperature occurring in any part of a winding insulation system, which is assumed to represent the thermal limitation for transformer loading. Following IEC60354 [6], for ON cooling, the HST under any load K is the sum of the ambient temperature, the top-oil temperature rise and and the hot spot rise above the top-oil temperature; and for OF cooling, the HST under any load K is equal to the sum of the ambient temperature, the bottom-oil temperature rise, the difference between the top-oil in the winding and the bottom-oil and the difference between the HST and the top-oil in the winding. In general, most of recorded HSTs come from previous heat run tests when the loading ratios are 0.5 and 1.0 in ONAN and OFAF conditions, respectively. Therefore, it is reasonable to calculate HSTs as the sum of the top-oil temperature and the hot-spot rise above the top-oil temperatures $\Delta\theta_{\mathrm{H}}$ by rearranging the IEC HST equations. Moreover, the recorded HSTs reached equilibria in manufacture heat run tests, so the use of the top-oil temperatures for the calculation of HSTs is applicable in most cases. Then, the HST can be calculated as:

$$\theta_{\mathrm{HST}} = \theta_{\mathrm{Oil-out}} + \theta_{\mathrm{a}} + \Delta\theta_{\mathrm{H}} = \mathrm{TOT} + \theta_{\mathrm{g}}K^{y}, \tag{4.17}$$

where θ_{g} is the hot-spot to top-oil gradient at the rated load current and TOT is the measured or simulated top-oil temperature rise. As for STEATM, the top-oil temperature rise $\theta_{\text{Oil-out}}$ is one of the model outputs, in order to obtain HSTs only the hot-spot rise is required in the form of $\theta_{\mathrm{g}}K^{y} = Hg_{\mathrm{r}}K^{y}$ [6].

4.8 Summary

In this chapter, the conventional IEC and IEEE thermal models are reviewed to provide the fundamental knowledge of transformer thermal modelling and ratings. Two equivalent heat circuits for modelling oil-immersed power transformers are developed, which are based on the principles of heat exchange and electrical circuit laws according to actual heat transfer mechanisms. The two thermoelectric analogy thermal models are established to calculate real-time temperatures of the main parts of ONAN/OFAF power transformers under various ambient and load conditions. Considering the operating regimes of transformer cooling systems and the actual measurements available, two temperature measurements, i.e. BOTs and TOTs, are chosen as model outputs, which can be subsequently used to calculate HSTs.

References

1. Transformers Committee of the IEEE Power Engineering Society (1991) IEEE guide for loading mineral oil-immersed transformer. IEEE Std. C57.91-1995, The Institute of Electrical and Electronics Engineers, Inc., New York
2. International Electrotechnical Commission (1993) IEC600762-power transformers—part 2: temperature rise, International Electrotechnical Commission Standard, Geneva
3. Galdi V, Ippolito L, Piccolo A, Vaccaro A (2000) Neural diagnostic system for transformer thermal overload protection. IEE Proc Electr Power Appl 47(5):415–421
4. Tang WH, Zeng H, Nuttall KI, Richardson ZJ, Simonson E, Wu QH (2000) Development of power transformer thermal models for oil temperature prediction. EvoWrokshops 2000, April 2000, Edinburgh, pp 195–204
5. Glenn S (2001) A fundamental approach to transformer thermal modelling. IEEE Trans Power Deliv 16(2):171–180
6. International Electrotechnical Commission (1991) IEC60354: loading guide for oil-immersed power transformers. International Electrotechnical Commission Standard, Geneva
7. Fourier (1992) Analytical theory of heat (translated by Freeman, A), Stechert, New York
8. Simonson JR (1981) Engineering heat transfer, The City University, London
9. Tang WH, Wu QH, Richardson ZJ (2002) Equivalent heat circuit based power transformer thermal model. IEE Proc Electr Power Appl 149(2):87–92
10. Alegi GL, Black WZ (1990) Real-time thermal model for an oil-immersed, forced-air cooled transformer. IEEE Trans Power Deliv 5(2):991–999
11. Wu QH, Ma JT (1995) Power system optimal reactive dispatch using evolutionary programming. IEEE Trans Power Syst 10(3):1243–1249
12. Tang WH, Wu QH, Richardson ZJ (2004) A simplified transformer thermal model based on thermal-electric analogy. IEEE Trans Power Deliv 19(3):1112–1119

Chapter 5
Thermal Model Parameter Identification and Verification Using Genetic Algorithm

Abstract In the previous chapter, two thermoelectric analogy thermal models have been developed to model thermal dynamics of oil-immersed transformers, i.e. a comprehensive thermoelectrical analogy thermal model and a simplified thermoelectrical analogy thermal model. For the problem of thermal parameter identification, a simple genetic algorithm is employed in this chapter to search global solutions of thermal model parameters using on-site measurements. Firstly, the parameter identification and verification of the comprehensive thermal model is presented. GA modelling results for the comprehensive model are compared with the historical heat run tests and the modelling results from an ANN model. For the simplified thermal model, a number of rapidly changing load scenarios are employed to verify the derived thermal parameters and finally an error analysis is given to demonstrate the practicability of the simplified thermal model.

5.1 Introduction

To verify the two thermoelectric analogy models developed in the preceding chapter, tests have been carried out using CTEATM and STEATM involving two large power transformers, i.e. SGT3A and SGT3B. The two equivalent heat circuit models are employed to calculate both the transient-state temperatures and the stationary equilibria of the two ONAN/OFAF cooled power transformers. For the problem of thermal parameter identification, an SGA in "The genetic algorithm and direct search toolbox of MATLAB" is employed to search global solutions for thermal parameters using on-site measurements.

W. H. Tang and Q. H. Wu, *Condition Monitoring and Assessment of Power Transformers Using Computational Intelligence*, Power Systems,
DOI: 10.1007/978-0-85729-052-6_5,

The transformer operation data provided by NG cover both the ONAN and OFAF cooling regimes, which are switched over automatically. During model calculations, transient-state thermal models actually fall into two groups, which are expressed by two sets of parameters, respectively. Therefore, the two sets of heat transfer parameters are used for the model optimisation and simulation, which reflect changes of thermal time constants due to different cooler operating conditions of a power transformer. The first set reflects the normal state relationship between the TOTs, BOTs, ambient temperatures and transformer load ratios, and the second set concerns temperature variations over a period of time as an inertial response to the function of an external cooler. In this chapter, five datasets are employed for model parameter identification and verification: dataset 1 is used for the parameter identification and verification of CTEATM; dataset 2 is involved in a comparison study between the SGA modelling and ANN modelling concerning CTEATM; and datasets 3–5 are used for the parameter identification and verification of STEATM.

5.2 Unit Conversion for Heat Equivalent Circuit Parameters

A common mathematical formulation of the two thermal models developed in Chap. 4 is shown below:

$$\mathbf{P} = \mathbf{G}\Theta + \mathbf{C}\,\mathrm{d}\Theta/\mathrm{d}t,$$

which is rearranged into the following format:

$$\mathrm{d}\Theta/\mathrm{d}t = (\mathbf{P} - \mathbf{G}\Theta)/\mathbf{C}.$$

Generally, the variable units used in the above equation are the international units, e.g. the unit related to time t, for the estimated loss $\mathbf{P}$, the heat conductance $\mathbf{G}$ and the heat capacitance $\mathbf{C}$, is second. However, the temperature data used in this research are sampled every minute, which are not in the unit of second due to a practical reason. As the unit of the input temperature data is in minutes, the unit of t of the item $\mathrm{d}\Theta/\mathrm{d}t$ is then in minutes. As all the other parameters' base units are defined with respect to second, it is therefore necessary to convert the parameter units in the above two equations to reflect the unit difference between model inputs and derived model thermal parameters, i.e. $\mathbf{P}$, $\mathbf{G}$ and $\mathbf{C}$.

Suppose the time unit of each parameter described in the equation below is in minutes:

$$\mathrm{d}\Theta/\mathrm{d}t^* = (\mathbf{P}^* - \mathbf{G}^*\Theta)/\mathbf{C}^*. \tag{5.1}$$

Regarding model calculations, the input temperature data are sampled every minute, which are fixed by an on-site sampling programme. Then, all the units of the other parameters should be converted into the units using 1 min as a sampling time interval. The unit conversion processes are introduced as follows. Firstly, the unit of t^* in Eq. 5.1 is minute as the input of $\boldsymbol{\Theta}$ is sampled in minutes; therefore, the units of all the other parameters should also be in minutes. Let us consider the unit for $\mathbf{P}^*$. According to the definition of power, the unit watt (W) is the power that gives rise to the production of energy at the rate of 1 J/s. As the interval of t^* is in minute, the unit of $\mathbf{P}^*$ in minute should be 1 W times 60. The unit of $\mathbf{C}$ is farad (F), the definition of which is the capacitance of a capacitor between the plates of which there appears a difference of potential of 1 V when it is charged by a quantity of electricity equal to 1 coulomb (1 C = 1 As), and so the unit of $\mathbf{C}^*$ in minutes should be 1 F times 60. In the same manner, the international unit of the conductance $\mathbf{G}$ in seconds is 1 S, which is equivalent to 1 A/V (i.e. 1 C/Vs); hence, when the unit of $\mathbf{G}$ is converted from second to minute, the unit of the conductance $\mathbf{G}^*$ in minute is 1 S times 60.

5.3 Fitness Function for Genetic Algorithm Optimisation

To implement an SGA for thermal parameter identification, a fitness function and other relevant parameters of an SGA should be predefined. Usually, the error between measured variables and model outputs is defined as fitness. As stated previously, the equivalent heat circuit models are established to calculate both the transient-state temperatures and the stationary equilibria of an ONAN/OFAF cooled power transformer. Considering the operation regimes of the cooling system and actual measurements available, two temperature measurements, BOT and TOT, are selected as training targets for GA learning. Thus, the fitness function in this particular study should contain at least two terms corresponding to the fitness of BOT and TOT, respectively. For each individual of a GA generation, its total fitness value, f, is calculated as follows:

$$\begin{aligned} f &= f_{\mathrm{bo}} + f_{\mathrm{to}} \\ &= \sum_{k=1}^{N} \Delta\theta'_{\mathrm{bo}}(k)^2 + \sum_{k=1}^{N} \Delta\theta'_{\mathrm{to}}(k)^2 \\ &= \sum_{k=1}^{N} (\theta_{\mathrm{bo}}(k) - \theta_{\mathrm{mbo}}(k))^2 + \sum_{k=1}^{N} (\theta_{\mathrm{to}}(k) - \theta_{\mathrm{mto}}(k))^2, \end{aligned} \tag{5.2}$$

where f_{bo} and f_{to} are the total errors of the BOT and TOT of the model, respectively, $\Delta\theta'_{\mathrm{bo}}(k)$ and $\Delta\theta'_{\mathrm{to}}(k)$ the errors between real measurements and model outputs under appropriate operation conditions (e.g. varying ambient temperatures and loading ratios), respectively, $\theta_{\mathrm{mbo}}(k)$ and $\theta_{\mathrm{mto}}(k)$ on-site measurements of BOT and TOT, respectively, N the total number of measurement groups, and $\theta_{\mathrm{bo}}(k)$ and $\theta_{\mathrm{to}}(k)$ the outputs of BOTs and TOTs from the two proposed thermal models.

5.4 Parameter Identification and Verification for the Comprehensive Thermal Model

5.4.1 Estimation of Heat Transfer Parameters

Normally, to obtain the values of thermal parameters (i.e. C and G), experiments should be conducted. As stated previously, for a large online power transformer, it is not practical to shut it down and do a heat run test. Additionally, such a test does not provide satisfactory results, as the thermal parameters cannot be determined precisely enough in one test due to the nonlinearity of transformer thermal dynamics change over time. To tackle this problem, evolutionary computation algorithms can be employed as an alternative approach to thermal parameter identification, which have been practised widely in engineering applications. For example, GA [1, 2, 3] has been employed by the authors as a search method to optimise model parameters based on real-world data for transformer modelling and power dispatch. Before running a GA program, the boundary of each parameter is to be estimated.

As known, based on the density, volume and specific heat capacity of a heat transfer path of a transformer, the approximate thermal capacitance per unit length is expressed by the following equation:

$$C = c_{\mathrm{p}}\rho v.$$

Each thermal capacitance is only related to one temperature node, as there is no coupling capacitance between the nodes in the equivalent heat circuits. Therefore, the thermal capacitance bound of each major component can be estimated. For instance, with regard to transformer SGT3A listed in Table 5.1, the mass of the whole volume of oil is 80.0 tons and c_{p} is approximately 1.8 kJ/kgK, then

$$C_{\mathrm{Oil}} = 1.8 \times 80.0 \times 1{,}000.0 = 144.0 \times 10^3\ \mathrm{kJ/K}.$$

Table 5.1 Rating of transformer SGT3A

Name plate rating	180.0 MVA
U_{HV} (kV)/U_{LV} (kV)	275.0/66.0 kV
Iron losses	128.3 kW
Copper losses (rated load)	848.0 kW
Stray losses	113.0 kW
Weight of core and windings	100.0 tons
Weight of oil	80.0 tons
TOT (half rated load)	41.3°C
Type of cooling	OFAF
Factory/year	Hackbridge/1961
Site	St. Johns Wood, London, UK

Whilst, the mass of the winding and core is 100.0 tons and c_p is approximately 0.5 kJ/kgK, then

$$C_{\text{Fe and Cu}} = 0.5 \times 100.0 \times 1{,}000.0 = 50.0 \times 10^3 \ \text{kJ/K}.$$

Consequently, the overall thermal capacitance of transformer SGT3A is roughly 195.0×10^3 kJ/K = $(144.0 \times 10^3 + 50.0 \times 10^3)$ kJ/K.

As for the nonlinear thermal conductances, they are related to many physical properties of a power transformer, such as actual geometrical size, temperature rise and parameters of thermal characteristics. It is thus very important to choose an appropriate mathematical form to represent this nonlinear feature. From experience obtained in this research, they are represented by a set of formulae in an exponential form:

$$G = a\Delta\theta^b.$$

In the above formula, G is an element of the heat conductance matrix $\mathbf{G}, \Delta\theta$ represents the temperature difference or temperature rise with regard to a studied object, and a and b are arguments to be identified by an SGA.

For example, in consideration of heat conductances, if the oil average temperature rise at equilibrium is 40 K under the rated load, the value of G_1' can be estimated approximately as

$$G_1' = \frac{P_{\text{all}}}{\Delta\theta} = \frac{848.0 + 128.3 + 113.0}{40.0} = 27.2\,\text{kW/K}.$$

Hence, we are able to further estimate a and b on the basis of the form of $G = a\Delta\theta^b$, and so do other heat conductances.

5.4.2 Parameter Identification Using Genetic Algorithm

As stated above, TOT and BOT are selected as model outputs of CTEATM. The measured inputs (dataset 1) are ambient temperatures θ_a, loading ratios K, TOTs $\theta_{\text{Oil-out}}$ and BOTs $\theta_{\text{Oil-in}}$, comprising 7,000 groups of measurements with a sample interval of 1 min. The model parameters are identified using an SGA based on 3,500 groups of real measurements, and then verified on the remaining 3,500 groups. The comparisons between the model outputs and real measurements are shown in Figs. 5.1 and 5.2, respectively, while real daily cyclic loading ratios and the cooler switching signals are displayed in Figs. 5.3 and 5.4, respectively.

For SGT3A with the ratings given in Table 5.1, the thermal model parameters obtained from the GA search are listed in Table 5.2, where the heat capacitance is in the unit of 10^3 kJ/K and the heat conductance in the unit of kW/K. With regard to GA parameters, the crossover probability p_c is 0.95 and the mutation probability p_m is 0.08. The size of GA population is selected as 60 after several trials. During

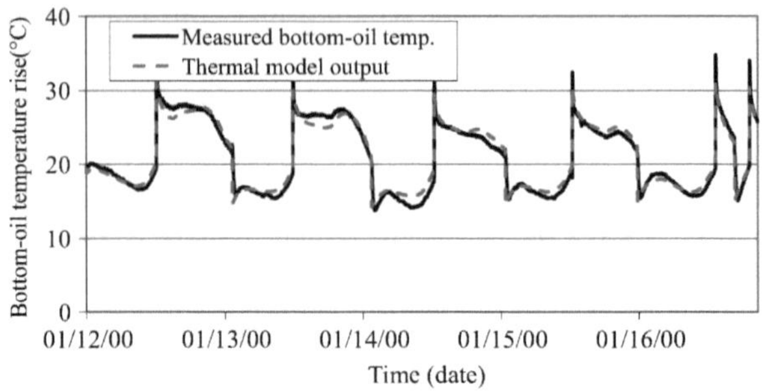

Fig. 5.1 Bottom-oil temperature of the CTEATM outputs compared with the on-site measurements (dataset 1) (maximum error, 1.9°C)

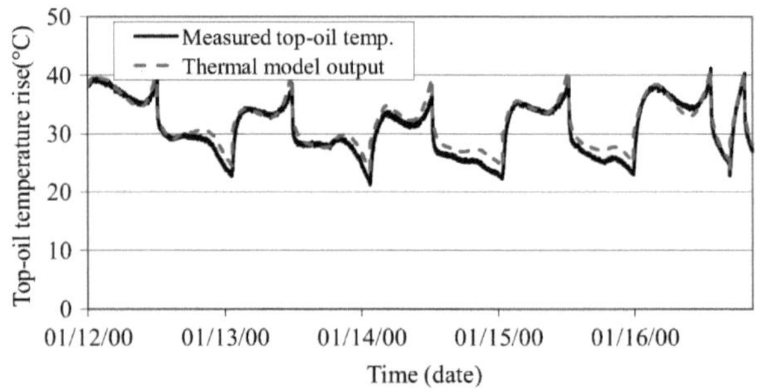

Fig. 5.2 Top-oil temperature of the CTEATM outputs compared with the on-site measurements (dataset 1) (maximum error, 2.1°C)

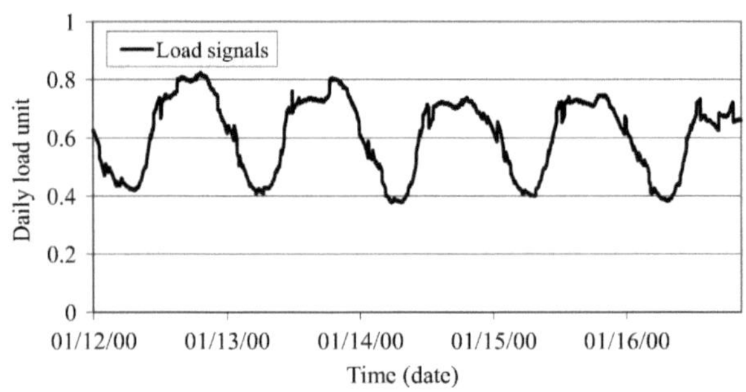

Fig. 5.3 Daily load (rated load, 1 unit) (dataset 1)

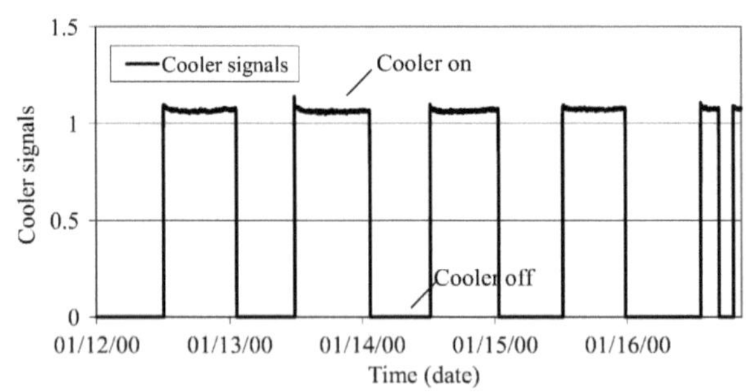

Fig. 5.4 Real-time cooler status (dataset 1)

Table 5.2 Thermal parameters of CTEATM for transformer SGT3A identified using SGA

Parameters	$G_1'(a)$	$G_1'(b)$	$G_2'(a)$	$G_2'(b)$	$G_3'(a)$
Cooler off	60.9	0.48	37.1	0.53	42.2
Cooler on	37.4	0.58	32.6	0.54	42.7
Parameters	$G_3'(b)$	$G_4'(a)$	$G_4'(b)$	$G_5'(a)$	$G_5'(b)$
Cooler off	0.21	5.8	0.41	2.8	0.67
Cooler on	0.13	133.3	0.51	4.2	0.55
Parameters	C_1'	C_2'	C_3'	C_4'	C_5'
Cooler off	28.8	31.9	102.5	3.1	0.13
Cooler on	27.6	35.7	49.3	3.6	1.7

the GA training, after 1,000 generations, the best solution indicates a lack of improvement. Therefore, the maximum number of generations is chosen as 1,000, which is set as the termination criterion of GA in this study. The block diagram of the SGA for obtaining the optimal thermal parameters is shown in Fig. 5.5. For model verification, the identified thermal parameters listed can be substituted into the differential equation (4.11) for temperature calculations with the online input data to produce continuous time sequences. During model calculations, the differential equation (4.11) is solved using the Runge–Kutta method in the time domain, by inputting the online data of ambient temperatures, cyclic loading ratios and cooler switching signals.

Satisfactory agreements between the model outputs and the real-time measurements are illustrated in Figs. 5.1 and 5.2, which suggest that CTEATM can reflect accurately the thermal dynamics of SGT3A. Notice that the data presented cover both the trained and untrained datasets and still show an accurate representation of actual thermal states with the untrained datasets, thereby the extrapolability of CTEATM is demonstrated to some extent. It can also be seen that the thermal model still holds a satisfactory performance even during the switching of coolers, which is of practical importance to determine transient thermodynamic states. In this regard, the conventional thermal models reported by IEEE or IEC cannot account for transient variations with different cooler states, which is a considerable drawback of the traditional thermal models.

5.4.3 Verification of Identified Thermal Parameters Against Factory Heat Run Tests

To further verify the identified thermal parameters listed in Table 5.2, which are obtained from the GA search, simulations are carried out under the same circumstances as historical factory heat run tests in accordance with IEC60076-2.

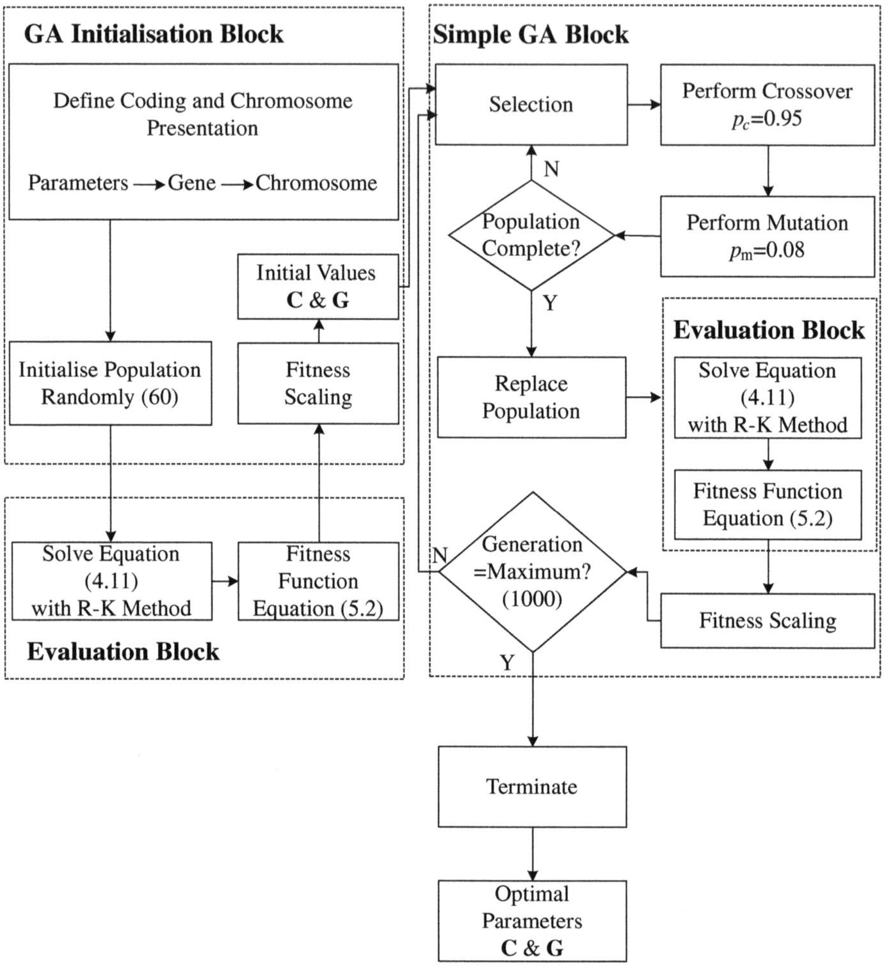

Fig. 5.5 Block diagram of SGA for searching optimal thermal parameters

Fig. 5.6 Bottom-oil temperature rise of the CTEATM outputs compared with the original heat run test (50% rated load, cooler off) (maximum error, 1.8°C)

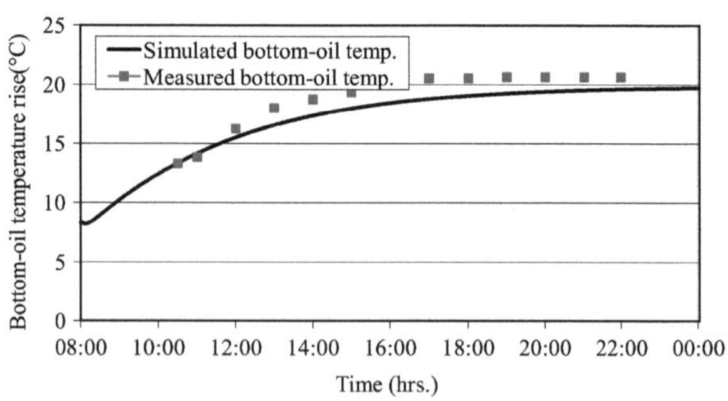

Table 5.3 Bottom-oil temperature rise (°C) at 50% rated load, cooler off of SGT3A

Time series	12:00	13:00	14:00	15:00	16:00
Tested	16.25	18.00	18.75	19.35	20.25
Simulated	15.51	16.56	17.36	17.97	18.44
Time series	17:00	18:00	19:00	20:00	21:00
Tested	20.55	20.55	20.65	20.65	20.65
Simulated	18.79	19.05	19.25	19.40	19.60

Table 5.4 Top-oil temperature rise (°C) at 50% rated load, cooler off of SGT3A

Time series	12:00	13:00	14:00	15:00	16:00
Tested	34.10	36.50	37.80	38.80	39.45
Simulated	31.07	33.28	34.97	36.27	37.25
Time series	17:00	18:00	19:00	20:00	21:00
Tested	39.85	41.05	41.05	41.05	41.05
Simulated	38.00	38.57	39.00	39.35	39.85

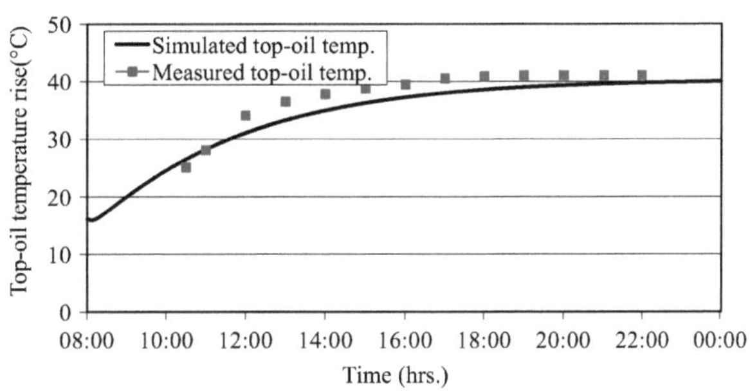

Fig. 5.7 Top-oil temperature rise of the CTEATM outputs compared with the original heat run test (50% rated load, cooler off) (maximum error, 2.8°C)

The following working conditions have been investigated: 50% rated load, cooler off, ambient temperature $\theta_a = 21.5°C$. Based on the thermal parameters identified from the GA search, the BOT and TOT of the model step responses are shown in Fig. 5.6, Table 5.3 and Fig. 5.7, Table 5.4, respectively, which are compared with the recorded historical test data under the same condition provided by the manufacturer of SGT3A [4]. It can be noticed that the data of the measured BOTs and TOTs recorded in the real heat run tests provided by NG are sparse. The oil temperature of the real heat run was significantly above the ambient temperature at the start of the tests and there was no information about any previous tests or loads. It is thus impossible to apply the same initial conditions to the simulations. To compare the temperatures from the real heat run tests with the results from model simulations, time-synchronisation has to been undertaken when drawing the two graphs together. The measured and simulated results are positioned with respect to each other along the two axes so as to minimise errors. These comments and actions also apply to the other heat run simulations in this chapter.

By comparing the simulated step responses and the measured responses in Figs. 5.6 and 5.7, respectively, it is evident that the simulated temperature rise is in close agreement with the actual measurements. Considering the sum of each capacitance as listed in Table 5.2 when the cooler is off, the summation value is 166.4×10^3 kJ/K, which is near the derived overall thermal capacitance of 195.0×10^3 kJ/K estimated in Sect. 5.4.1. Moreover, the time constants of the historical heat run tests, observed from the temperature–time graphs in Figs. 5.6 and 5.7, approach closely those based on the GA search determined by $\mathbf{C}'$ and $\mathbf{G}'$. As a consequence, the identified thermal conductances are also reliable, while the thermal capacitances are sufficiently accurate.

Nevertheless, the deviations, among the experiment simulations, the on-site measurements and the factory tests indicate that the parameters, obtained from the GA search should be re-evaluated carefully with regard to actual situations. To improve the thermal model performance, some further information is required, such as tap position signals, HV and LV winding current ratios, etc. In the meantime, a further study should be undertaken to discover the interrelationships between the model parameters and responses with respect to various fault scenarios. From the results obtained from CTEATM, it is deduced that:

1. The simulation of CTEATM has a good agreement with the on-site data based on two sets of parameters identified by the SGA.
2. CTEATM can predict transformer operating temperatures for new load scenarios, which does not require real TOT and BOT measurements once the thermal model parameters are identified. Thus, load ratings can be calculated continuously in real time.
3. CTEATM also allows simulations of specific and interesting load cases, under emergency conditions or overload requirements, e.g. to determine how long a specific load could be permitted, while with the maximum efficiency and the minimum thermal risk.
4. In addition, as the model has a clear physical meaning, it possesses great practicability for accurate temperature calculations.

5.4.4 Comparison between Modelling Results of Artificial Neural Network and Genetic Algorithm

ANN is based on the theory of biological nervous systems and involves the selection of input, output, network topology and weighted connections of network's nodes [5]. In general, neural networks can be adjusted or trained, so that a particular input leads to a specific target output. Batch training of ANNs proceeds by making weight and bias changes based on an entire batch of input vectors. ANNs have been employed widely to perform complex functions in various fields of applications including pattern recognition, classification and control systems.

In this study, a recursive mapping ANN in "The neural network toolbox of MATLAB" is configured to describe transformer thermal behaviours for predicting the BOTs and TOTs of SGT3A. The network's input vectors consist of the load ratio K, the cooler status c and the ambient temperature θ_a. The network's outputs are the BOTs or TOTs of SGT3A, and all the neuron functions are chosen to be the sigmoid function except the output neuron, which employs a pure linear function. Two $7 \times 9 \times 1$ recursive form ANNs are used to mimic the inertia and performance of BOT and TOT, respectively, by selecting the ANN inputs as current and previous samples:

$$[K(t-1),\ K(t-2),\ K(t-3),\ \theta_a(t-1),\ \theta_a(t-2),\ \theta_a(t-3),\ c(t)]$$

where t is the current sample time that can be shifted to the nth ($n = 1$, 2 and 3) earlier time.

During the ANN modelling, with dataset 2 (7,500 groups of on-site measurements), 70% of the total data are used to train the ANN network and the remaining 30% are reserved for evaluating the performance of the ANN modelling within the untrained range. The training epochs are 1,500. The actual measurements of dataset 2 are shown in Figs. 5.8 and 5.9, for loading ratios and cooler signals, respectively. The results obtained from the ANN recursive models are shown in Figs. 5.10 and 5.11. In comparison, CTEATM is also used to calculate the BOTs

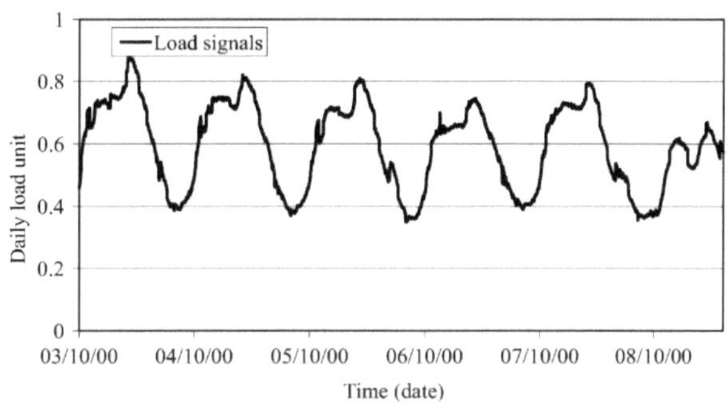

Fig. 5.8 Daily load (rated load, 1 unit) (dataset 2)

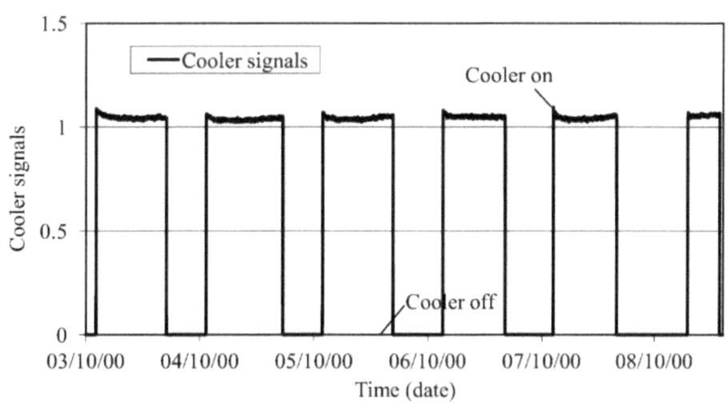

Fig. 5.9 Real-time cooler status (dataset 2)

and TOTs using the thermal parameters in Table 5.2 for dataset 2, and the outputs of CTEATM are illustrated in Figs. 5.12 and 5.13.

It can be observed that, when the recursive ANN mapping is used, the ANN model only represents the input and output relationship accurately within the range covered by the training data. Outside this range, the model response and measurements do not agree with each other satisfactorily, especially under rapid variations of load. This results only demonstrate the possibility of one step prediction of transformer temperatures using ANNs. Moreover, the ANN modelling does not possess any physical meaning, so its extrapolability is very limited. It is also noticed that satisfactory agreements between the CTEATM outputs and the actual measurements using the GA identified parameters are illustrated in Figs. 5.12 and 5.13, which suggest that CTEATM is able to reflect accurately the

Fig. 5.10 Bottom-oil temperature rise of the ANN model outputs compared with the on-site measurements (dataset 2) (maximum error, 8.9°C)

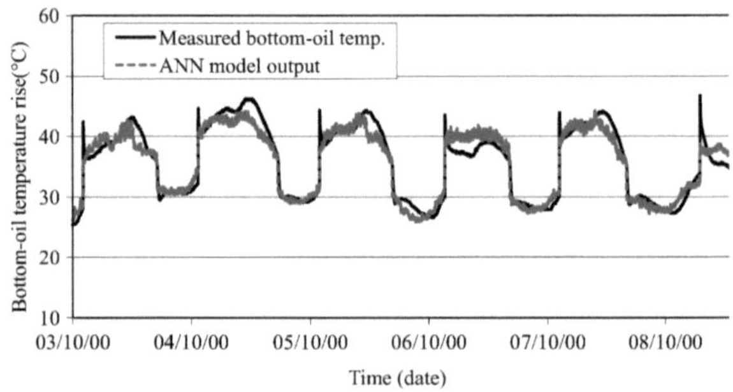

Fig. 5.11 Top-oil temperature rise of the ANN model outputs compared with the on-site measurements (dataset 2) (maximum error, 2.5°C)

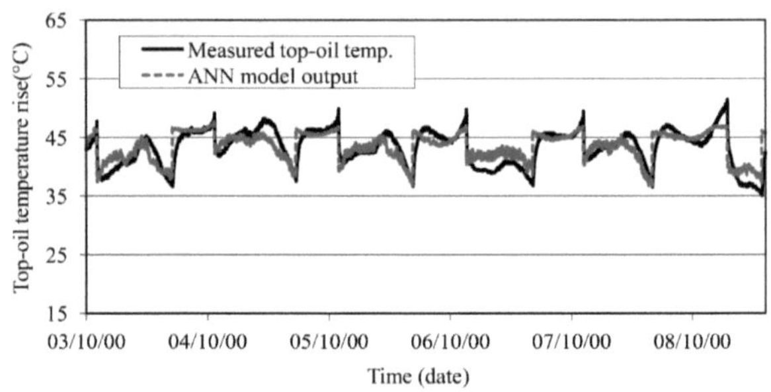

Fig. 5.12 Bottom-oil temperature rise of the CTEATM outputs compared with the on-site measurements (dataset 2) (maximum error, 1.4°C)

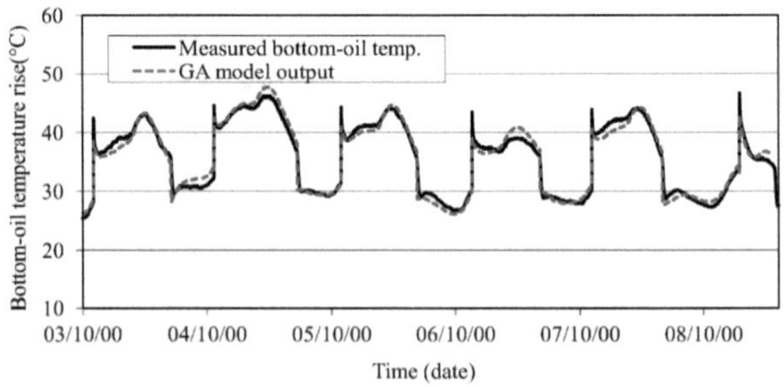

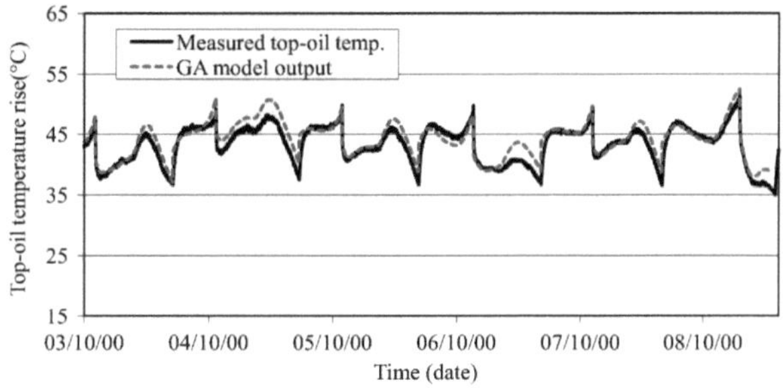

Fig. 5.13 Top-oil temperature rise of the CTEATM outputs compared with the on-site measurements (dataset 2) (maximum error, 1.3°C)

thermal dynamics of an oil-immersed power transformer. Moreover, as CTEATM has a clear physical meaning compared with the black-box technique of the ANN modelling, it is more useful in practice.

5.5 Parameter Identification and Verification for the Simplified Thermal Model

Regarding STEATM, the data employed for modelling also cover both the ONAN and OFAF operation regimes switched over automatically, which are sampled from a sister transformer of SGT3A as SGT3B. Three datasets are used, i.e. dataset 3, dataset 4 and dataset 5. Dataset 3 is employed in the parameter identification for STEATM and model verifications are carried out using datasets 4 and 5.

5.5.1 Identification of Parameters Using Genetic Algorithm

As for STEATM, the TOT and BOT are also selected as the model outputs. The measured inputs of each dataset consist of ambient temperatures θ_a, loading ratios K, TOTs ($\theta_{\text{Oil-out}}$) and BOTs ($\theta_{\text{Oil-in}}$), comprising 7,000 groups of measurements with a sample interval of 1 min. As the variations of θ_a do not affect the BOT and TOT instantly due to the thermal lagging, the model inputs of θ_a are a moving average of 5 h from field experience that are then fed into STEATM during calculations. The STEATM parameters are identified using the SGA based on 3,500 groups of real measurements, and then verified on the remaining 3,500 groups, noting that both the groups are extracted from dataset 3. During model calculations, the differential equation (4.15) is solved with the Runge–Kutta method in the time domain, by inputting the real data of ambient temperatures, cyclic loadings and cooler switching signals. For GA learning, the crossover probability p_c is 0.95 and the mutation probability p_m is 0.08. The size of the GA population is 100 and the maximum number of generations is 1,000.

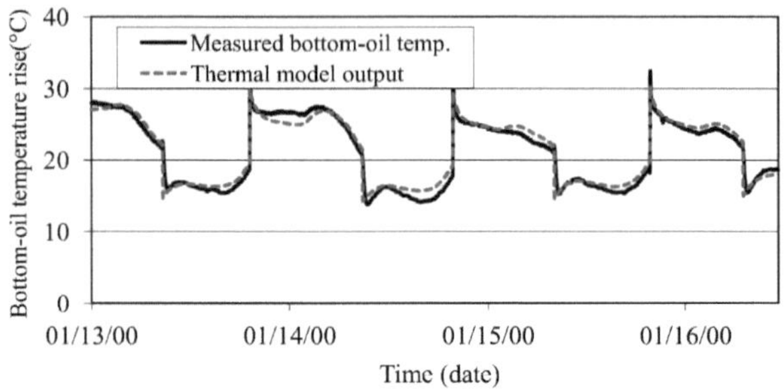

Fig. 5.14 Bottom-oil temperature rise of the STEATM outputs compared with the on-site measurements (dataset 3) (maximum error, 1.8°C)

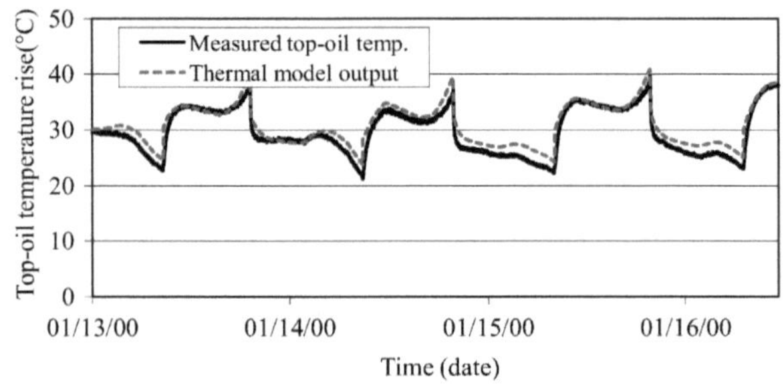

Fig. 5.15 Top-oil temperature rise of the STEATM outputs compared with the on-site measurements (dataset 3) (maximum error, 0.68°C)

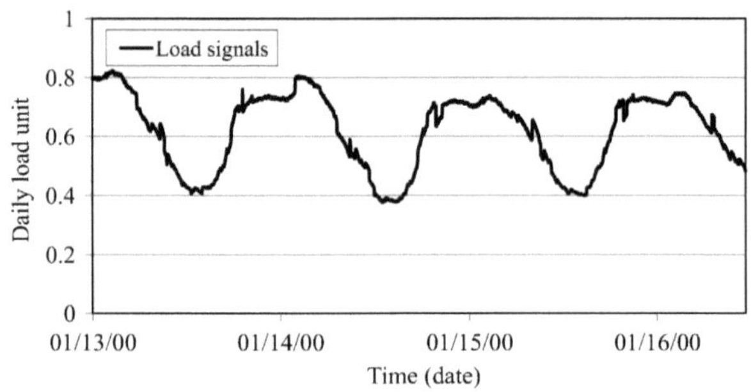

Fig. 5.16 Daily load (rated load, 1 unit) (dataset 3)

The comparisons between the model outputs and real measurements are shown in Figs. 5.14 and 5.15, respectively, while the real daily cyclic loading and the cooler switching signals are sketched in Figs. 5.16 and 5.17, respectively. Regarding SGT3B, the thermal model parameters obtained with a GA search are listed in Table 5.5, where the heat capacitance is in the unit of 10^3 kJ/K and the heat conductance in the unit of kW/K.

Satisfactory agreements between the model outputs and the on-site measurements are illustrated in Figs. 5.14 and 5.15, which suggest that STEATM can describe the thermal behaviours of the oil-immersed power transformer accurately under rapidly changing operation conditions. Notice that the data presented cover both the trained and untrained groups, which still give an accurate representation of transformer temperatures regarding the untrained groups.

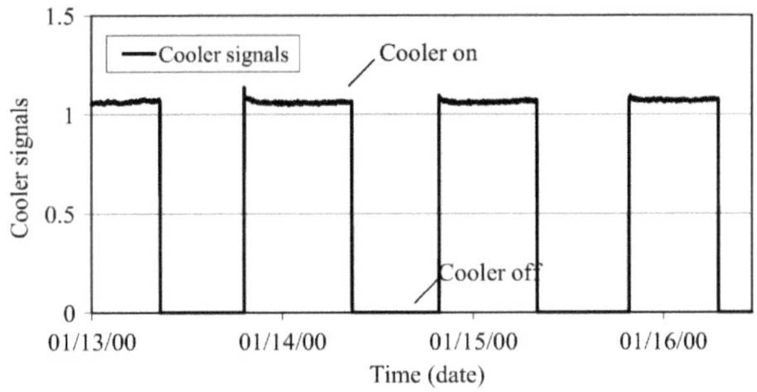

Fig. 5.17 Real-time cooler status (dataset 3)

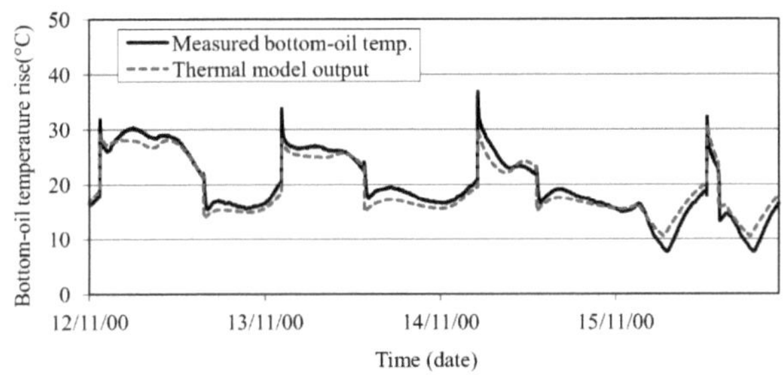

Fig. 5.18 Bottom-oil temperature rise of the STEATM outputs compared with the on-site measurements (dataset 4) (maximum error, 2.2°C)

Table 5.5 Parameters of STEATM for SGT3B identified with the SGA

Parameters	G_1(a)	G_1(b)	G_2(a)	G_2(b)	G_3(a)
Cooler off	49.2	0.34	3.64	0.54	3.59
Cooler on	38.5	0.24	126.3	0.52	4.08
Parameters	G_3(b)	C_1	C_2	C_3	
Cooler off	0.56	189.1	3.65	2.04	
Cooler on	0.54	123.4	3.07	1.93	

5.5.2 Verification of Derived Parameters with Rapidly Changing Loads

To verify the derived STEATM parameters under other operation conditions, another two working scenarios, sampled from the same in-service power transformer SGT3B, are employed for a further verification:

1. Dataset 4: loading conditions with intermittent switching off of SGT3B, which also accompanies with switches of the coolers as shown in Figs. 5.20 and 5.21.
2. Dataset 5: long-time small loading ratios, while coolers are off as shown in Figs. 5.24 and 5.25.

For simulations using STEATM, the thermal parameters listed in Table 5.5 are substituted into the differential equation (4.15) for model calculations with dataset

4 and dataset 5 to produce continuous time sequences under the two rare operation scenarios. Whereas, the differential equation (4.15) is solved with the Runge–Kutta method using the same parameters obtained from the GA optimisation based on dataset 3, involving the recorded data of ambient temperatures, loading ratios and cooler switching signals.

For dataset 4, Figs. 5.18 and 5.19 illustrate the measured and calculated temperature rise curves, respectively, while the real daily cyclic loading and cooler switching signals are displayed in Figs. 5.20 and 5.21, respectively. Considering dataset 5, the comparisons between the model outputs and real measurements are shown in Figs. 5.22 and 5.23, respectively, while the real daily loading ratios and cooler switching signals are displayed in Figs. 5.24 and 5.25, respectively.

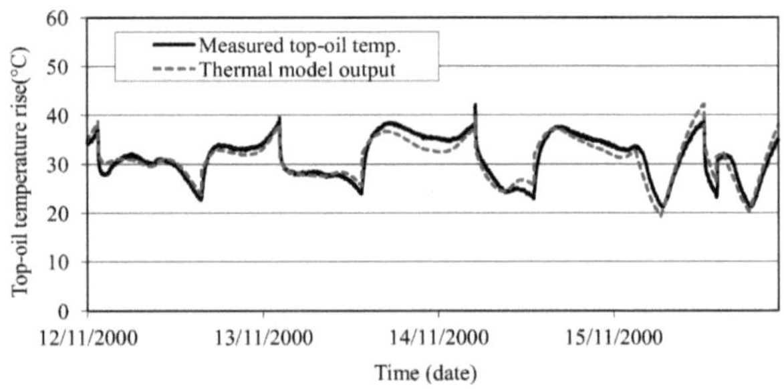

Fig. 5.19 Top-oil temperature rise of the STEATM outputs compared with the on-site measurements (dataset 4) (maximum error, 3.3°C)

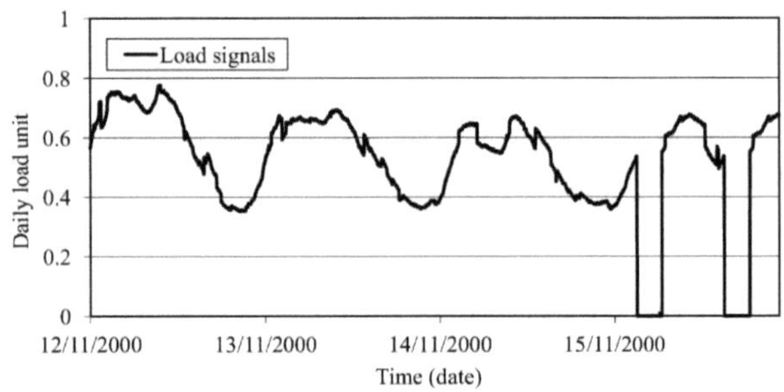

Fig. 5.20 Daily load (rated load, 1 unit) (dataset 4)

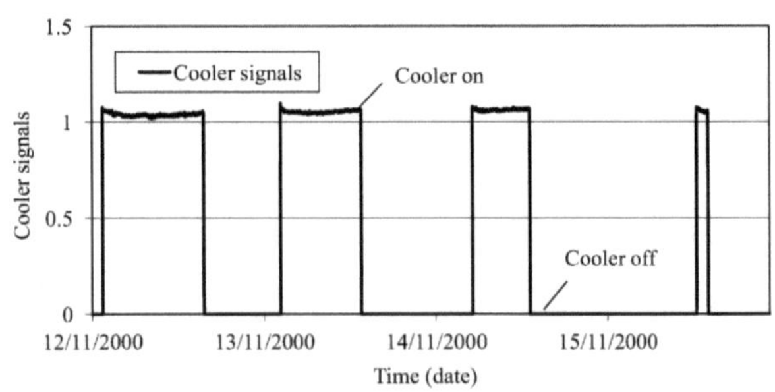

Fig. 5.21 Real-time cooler status (dataset 4)

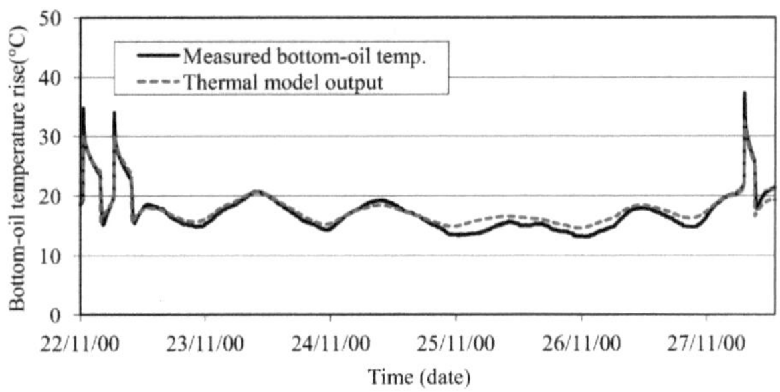

Fig. 5.22 Bottom-oil temperature rise of the STEATM outputs compared with the on-site measurements (dataset 5) (maximum error, 2.0°C)

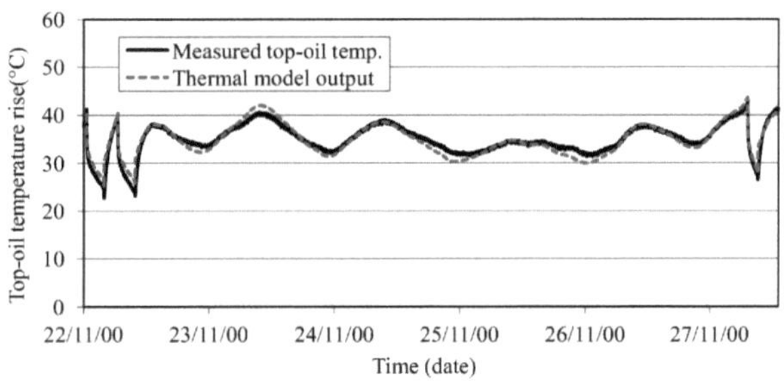

Fig. 5.23 Top-oil temperature rise of the STEATM outputs compared with the on-site measurements (dataset 5) (maximum error, 3.3°C)

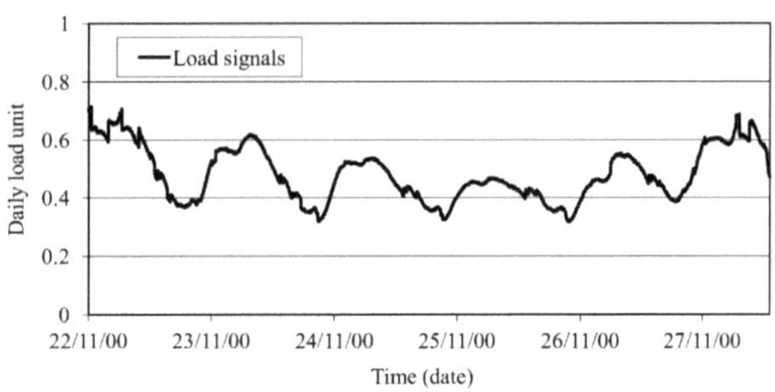

Fig. 5.24 Daily load (rated load, 1 unit) (dataset 5)

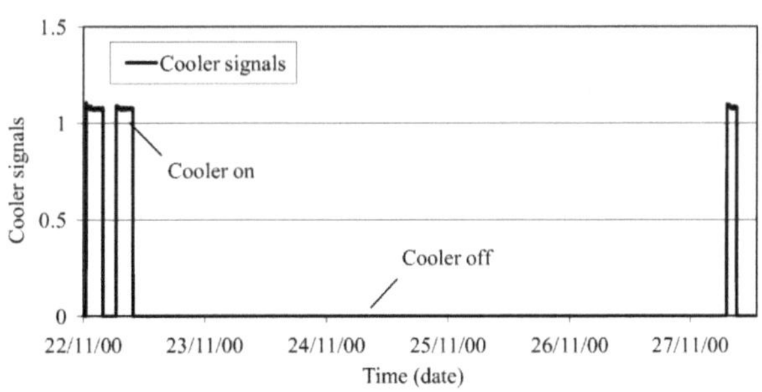

Fig. 5.25 Real-time cooler status (dataset 5)

It can be noted that STEATM still produces accurate results with the two selected loading scenarios in a practical sense, which could further demonstrate the usability of STEATM to some extent with the two sets of thermal parameters derived using a GA.

5.5.3 Simulations of Step Responses Compared with Factory Heat Run

For the purpose of further validating the STEATM parameters in Table 5.5 under steady states, several simulations using STEATM have been carried out under the same circumstances as the historical factory heat run tests in accordance with IEC60076-2. Firstly, the following working conditions have been investigated: 50% rated load, coolers off, $\theta_a = 22.0°C$. The BOT and TOT of the model step responses, based on the thermal parameters derived from the GA search, are shown in Fig. 5.26, Table 5.6 and Fig. 5.27, Table 5.7 respectively, compared with the recorded historical test data under the same condition provided by the manufacturer of SGT3B.

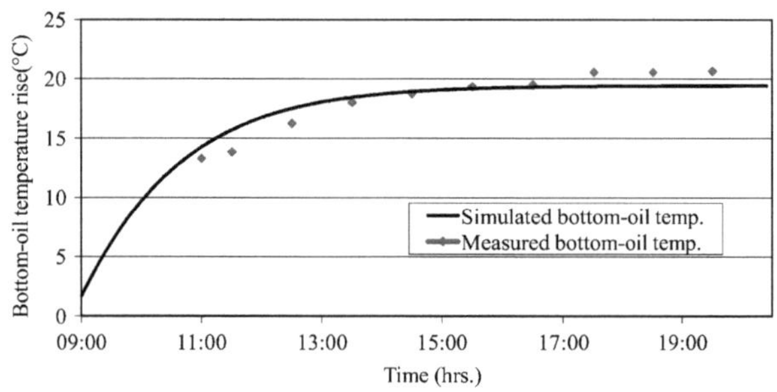

Fig. 5.26 Bottom-oil temperature rise of the STEATM outputs compared with the original heat run test (50% rated load, cooler off) (maximum error, 1.6°C)

Table 5.6 Bottom-oil temperature rise (°C) at 50% rated load, cooler off of SGT3B

	11:00	11:30	12:30	13:30	14:30
Tested	13.3	13.9	16.3	18.0	18.8
Simulated	14.3	15.5	17.4	18.5	18.8
	15:30	16:30	17:30	18:30	
Tested	19.4	20.3	20.6	20.7	
Simulated	19.2	19.3	19.4	19.4	

Table 5.7 Top-oil temperature rise (°C) at 50% rated load, cooler off of SGT3B

	11:00	11:30	12:30	13:30	14:30
Tested	25.2	29.1	35.1	37.5	38.8
Simulated	26.6	29.9	33.9	36.1	37.2
	15:30	16:30	17:30	18:30	
Tested	39.8	40.5	40.9	41.1	
Simulated	37.7	38.1	38.2	38.3	

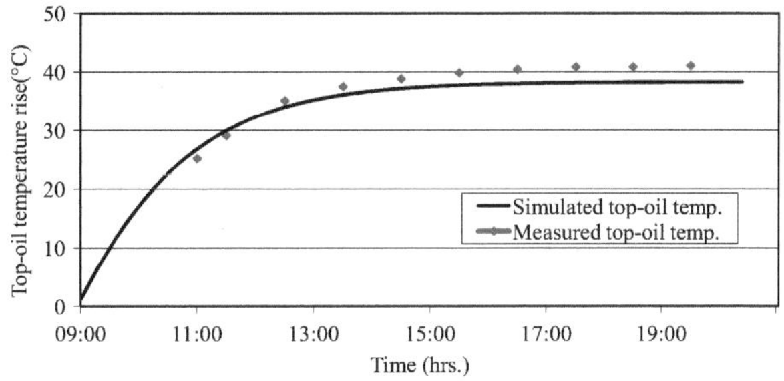

Fig. 5.27 Top-oil temperature rise of the STEATM outputs compared with the original heat run test (50% rated load, cooler off) (maximum error, 2.8°C)

Table 5.8 Bottom-oil temperature rise (°C) at 100% rated load, cooler on of SGT3B

	14:00	15:00	16:00	17:00	18:00
Tested	17.6	24.3	28.8	30.8	32.1
Simulated	19.1	26.9	31.1	33.2	34.3
	19:00	19:30	20:00	20:30	21:30
Tested	33.4	33.7	33.8	34.2	34.4
Simulated	34.9	35.0	35.1	35.2	35.3

Table 5.9 Top-oil temperature rise (°C) at 100% rated load, cooler on of SGT3B

	13:00	14:00	15:00	16:00	17:00
Tested	3.5	19.7	27.0	31.9	34.3
Simulated	5.5	20.2	29.0	33.7	36.1
	18:00	19:00	19:30	20:00	20:30
Tested	35.9	37.2	37.7	38.0	38.2
Simulated	37.3	37.9	38.1	38.2	38.3

Another full load condition has also been investigated: 100% rated load, coolers on, $\theta_a = 21.6$°C. The BOT and TOT of model step responses, based on the thermal parameters derived from the GA search, are presented in Fig. 5.28, Table 5.8 and Fig. 5.29, Table 5.9 respectively, which are again compared with the recorded historical test data under the same conditions.

Comparing the simulated step responses with the recorded historical test responses in Figs. 5.26, 5.27, 5.28 and 5.29, it is evident that the simulated temperature rise curves are in close agreement with the actual factory measurements. It can also be found in Tables 5.6, 5.7, 5.8 and 5.9 that the simulated final equilibria of the two operation scenarios are close to the recorded measurements. The time constants of the two pairs of curves, which are observed from the temperature–time graphs in Figs. 5.26, 5.27, 5.28 and 5.29 and determined by **C** and **G**, are around 2.7 and 1.2 h, respectively, extracted using exponential approximations, which are also close to the historical test measurements. Considering the

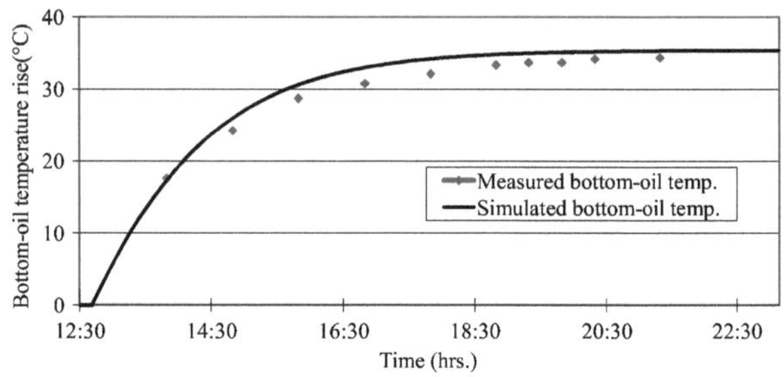

Fig. 5.28 Bottom-oil temperature rise of the STEATM outputs compared with the original heat run test (100% rated load, cooler on) (maximum error, 2.3°C)

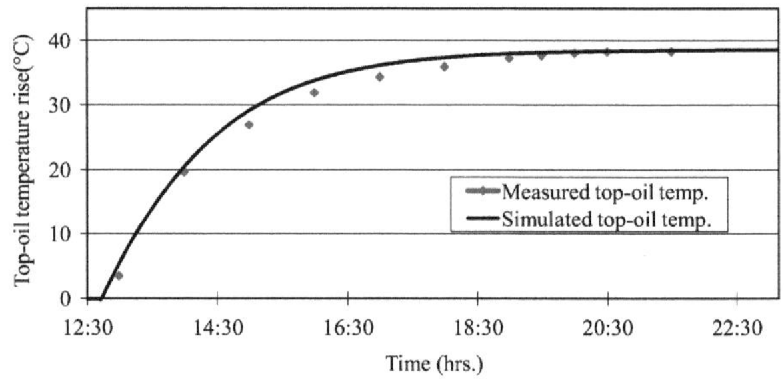

Fig. 5.29 Top-oil temperature rise of the STEATM outputs compared with the original heat run test (100% rated load, cooler on) (maximum error, 2.0°C)

summation of the heat capacitances C_1, C_2 and C_3, regarding Table 5.5 when the coolers are off, the value is 194.8×10^3 kJ/K, which is near the estimated value C_{all} of 195.0×10^3 kJ/K in Sect. 5.4.1. As a consequence, the thermal conductances based on the GA search are also reliable, while the thermal capacitances are sufficiently accurate. The simulation results of the full load condition are satisfactory, which can verify the extrapolability of the thermal model. It should be noticed that the highest training load ratio is under 80%, and during the simulation under the full load condition the loading ratio is 100%.

5.5.4 Hot-Spot Temperature Calculation

To derive the equilibrium HSTs under the conditions of the two heat run tests in Sect. 5.5.3, the hot-spot temperatures are calculated with Eq. (4.17) by setting y, θ_g (ONAN) and θ_g (OFAF) to 1.6, 26.0°C and 22.0°C, respectively [6].

Thus, under the ONAN, half load condition, the equilibrium of HST is calculated based on the model outputs in Table 5.7:

$$\begin{aligned}\theta_{\text{HST}} &= \theta_{\text{Oil-out}} + \theta_a + \theta_g K^y \\ &= 38.3 + 22.0 + 26.0(0.5)^{1.6} = 68.9°\text{C},\end{aligned}$$

and under the OFAF, full load condition, the equilibrium of HST is obtained on the basis of the model outputs in Table 5.9:

$$\theta_{\mathrm{HST}} = \theta_{\mathrm{Oil-out}} + \theta_{\mathrm{a}} + \theta_{\mathrm{g}} K^{y}$$
$$= 38.3 + 21.6 + 22.0(1.0)^{1.6} = 81.9°\mathrm{C}.$$

During the two historical heat run tests, the HSTs were recorded by a WTI. The final recorded equilibria of HSTs were 70.5°C and 82.0°C for the ONAN, half load condition and the OFAF, rated load condition, respectively. On comparing the calculated HSTs with the recorded values, they are very close to each other in both the scenarios.

5.5.5 *Error Analysis*

By defining the average error of BOT as $\sqrt{f_{\mathrm{bo}}/N}$ and the average error of TOT as $\sqrt{f_{\mathrm{to}}/N}$, the errors of the simulations discussed in this section are listed in Table 5.10. It can be observed from Table 5.10 that the largest average error of all calculations illustrated is <2°C, which is close to the results derived from CTEATM in the previous section. This shows that further simplification of the thermal model still holds good accuracies with a clear physical meaning, yet with less parameters to be identified.

5.6 Summary

In this chapter, tests have been carried out on two oil-immersed power transformers (SGT3A and SGT3B) using CTEATM and STEATM, respectively, including a detailed discussion and validation of the two developed thermal models. The robustness of GAs has been demonstrated through a number of case studies using the two models. The model parameters can be identified by a GA search based on the on-site measurements of ambient temperatures and load currents. A comparison study between the ANN modelling and GA modelling is also presented concerning CTEATM. From the responses of the two thermoelectric analogy models, it can be deduced that the thermoelectric analogy models can represent accurately real thermal dynamics of oil-immersed power transformers.

Table 5.10 Average errors of simulations using STEATM (°C)

	Scenario of dataset 4	Scenario of dataset 5	Heat run simulation ONAN, 50%	Heat run simulation OFAF, 100%
$\sqrt{f_{\mathrm{bo}}/N}$	0.85	1.69	1.01	1.78
$\sqrt{f_{\mathrm{to}}/N}$	1.32	1.74	1.93	1.32

References

1. Wu QH, Cao YJ, Wen JY (1998) Optimal reactive power dispatch using an adaptive genetic algorithm. Int J Electr Power Energy Syst 20(8):563–569
2. Tang WH, Zeng H, Nuttall KI, Richardson ZJ, Simonson E, Wu QH (2000) Development of power transformer thermal models for oil temperature prediction. EvoWorkshops 2000, April 2000, Edinburgh, Scotland, UK, pp 195–204
3. Tang WH, Wu QH, Richardson ZJ (2004) A simplified transformer thermal model based on thermal-electric analogy. IEEE Trans Power Deliv 19(3):1112–1119
4. The British Electric Transformer Ltd: (1967) Transformer test report of St. Johns Wood S.8. The British electric Transformer Ltd, Hayes
5. Zaman MR (1998) Experimental testing of the artificial neural network based protection of power transformers. IEEE Trans Power Deliv 13(2):510–517
6. International Electrotechnical Commission: (1991) IEC60354: loading guide for oil-immersed power transformers. International Electrotechnical Commission Standard, Geneva

Chapter 6
Transformer Condition Assessment Using Dissolved Gas Analysis

Abstract The dissolved gas analysis (DGA) of transformers can provide an insight view related to thermal and electrical stresses during operations of oil-immersed power transformers. DGA has been practised widely to detect incipient transformer faults and can therefore prevent any further damage to transformers. This chapter focuses on a literature review concerning conventional DGA techniques, as well as the recent advance in DGA diagnostic techniques. Firstly the gas evolution in a transformer is introduced. Various conventional DGA diagnosis methods are then presented, which are usually combined to give a comprehensive view of internal characteristics of transformers, such as the Rogers ratio method, the key gas method, the gassing ratio method, etc. Finally, a brief introduction to the diagnostic techniques using CI for DGA are presented.

6.1 Introduction

As known, oil-filled power transformers are subject to electrical and thermal stresses. The two stresses could break down insulation materials and release gaseous decomposition products [1], although all oil-filled transformers generate a small quantity of gas, particularly carbon monoxide (CO) and carbon dioxide (CO_2), to some extent at normal operation conditions. Overheating, corona (partial discharge) and arcing are the three primary causes of fault related gas generation, and such internal faults in oil produce gaseous byproducts, including hydrogen (H_2), methane (CH_4), acetylene (C_2H_2), ethylene (C_2H_4) and ethane (C_2H_6). When the cellulose is involved, a fault may produce CH_4, H_2, CO, CO_2, etc. Each of these types of faults produces certain gases that are generally combustible. The total of all combustible gases (TCG) with increases of gas generating rates may indicate the existence of any one or a combination of thermal, electrical or corona faults. Certain combinations of each of separate gases, called key gases, are unique

W. H. Tang and Q. H. Wu, *Condition Monitoring and Assessment of Power Transformers Using Computational Intelligence*, Power Systems,
DOI: 10.1007/978-0-85729-052-6_6,

for different fault temperatures and the ratios of certain key gas concentrations are indicative to fault types.

The dissolved gas analysis (DGA) has been a widely utilised and powerful tool to detect incipient faults in oil-filled power transformers [2–5]. The traditional practice of diagnosing transformer conditions through DGA is carried out off-line by manually extracting a sample of transformer insulation oil (by syringe), sending it to a laboratory, and waiting for diagnosis results. By applying DGA techniques on an oil sample, dissolved gases can be quantified. The concentration and the relation of individual gases can predict whether a fault has occurred and what type it is likely to be. Over the last four decades, DGA and its interpretation have become a popular and reliable tool for assessing conditions of oil-filled transformers and other oil-filled electrical equipment.

6.2 Fundamental of Dissolved Gas Analysis

Different patterns of gases are generated due to different intensities of energy dissipated by various faults in a power transformer. Totally or partially dissolved into the insulation oil, the gases present in an oil sample make it possible to determine the nature of a fault by gas types and their concentrations.

The most widely used dissolved gas extraction process is to get an oil sample through a sampling valve and inject it into an oil–gas extractor for analysis, using chromatography, mass spectrometry, infrared analytical methods, and so forth. After extraction and analysis, types of different gases and each concentration are determined, which can be compared with gas analysis records in a laboratory gathered over decades, followed by an evaluation of impact of a fault on the serviceability of a power transformer. Once a suspicious gas presence is detected, further inspections should be carried out to identify the species and locations of faults, such as tests of no-load characteristics of winding DC resistance, insulation, partial discharge or humidity content measurements, etc.

Various diagnostic schemes have been developed for DGA interpretation. These methods attempt to map the relations between gases and fault conditions, some of which are obvious and some of which may not be apparent. For instance, these criteria include the key gas method and the gas ratio method based on variations in gassing characteristics with temperatures to which materials are subjected. The comparisons of these DGA interpretation schemes indicate a large variety of ratios and typical values for individual gas concentrations between different DGA diagnostic schemes.

6.2.1 Gas Evolution in a Transformer

Faults in a transformer sometimes lead to the degradation of insulation materials and oil. During this degradation, gaseous products are formed and dissolved in

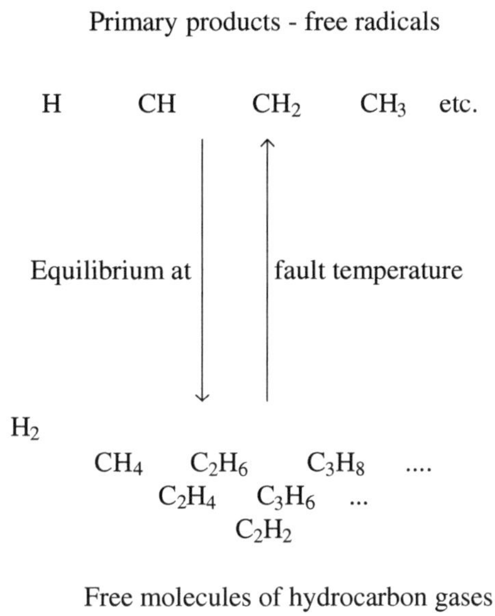

Fig. 6.1 Free radicals resulting from heating of mineral oil

the oil, namely H_2, CO, CO_2, CH_4, C_2H_6, C_2H_4, C_2H_2, etc. If a certain level is exceeded, gas bubbles arise and oil-filled transformers are subject to electrical and thermal stresses, which may break down insulation materials and release gaseous decomposition products. Evaluation procedures for DGA have been implemented widely based upon the guidelines recommended by IEC [2], IEEE [3] and CIGRE [4].

The immediate effect of the breakdown of hydrocarbon molecules as a result of the energy of a fault is to create free radicals as indicated in Fig. 6.1[1]. These subsequently recombine to produce low molecular weight hydrocarbon gases. This recombination process is largely determined by operation temperatures, but also influenced by other factors. The result is that the pattern of gases appearing in the oil has a form as shown in Fig. 6.2. For the lowest temperature faults both CH_4 and H_2 are generated, with CH_4 being predominant. As the temperature of a fault increases, C_2H_6 starts to be evolved and CH_4 is reduced, so that the C_2H_6/CH_4 ratio becomes predominant. At still higher temperatures the rate of C_2H_6 evolution is reduced and C_2H_4 production commences and soon outweights the proportion of C_2H_6. Finally, at very high temperatures C_2H_4 puts in an appearance and as the temperature increases still further it becomes the most predominant gas. It is noted that no temperature scale is indicated along the temperature axis of Fig. 6.2, and the graph is subdivided into types of faults. The areas include normal operating temperatures go up to about 140°C, hot spots extend to around 250°C and high-temperature thermal faults to about 1000°C. Peak C_2H_4 evolution occurs at about 700°C [1].

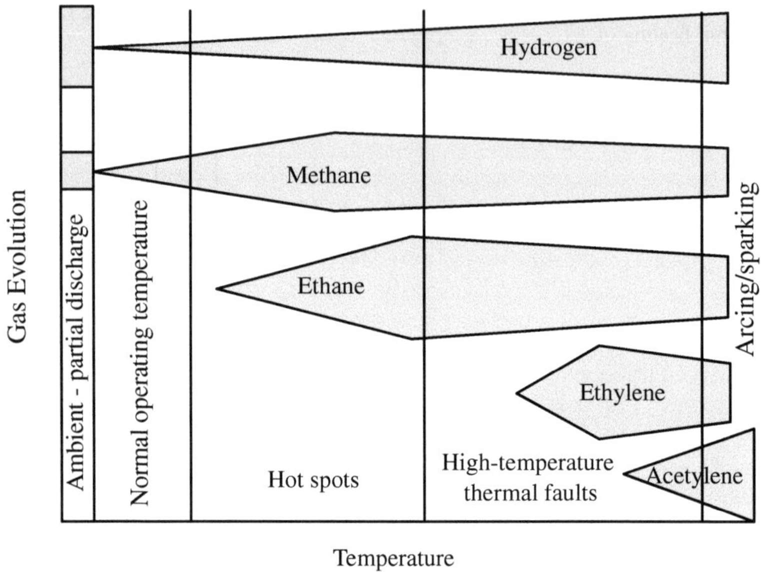

Fig. 6.2 Hydrocarbon gas evolution in transformer oil against temperature

6.2.2 Key Gas Method

The key gas method relates key gases to fault types and attempts to detect four fault types [3], including overheating of oil, overheating of cellulose, corona (partial discharge) and arcing, based on key gas concentrations (C_2H_4, CO, H_2, C_2H_2) expressed in ppm (part per million).

1. Overheating of oil: Decomposition products include C_2H_4 and CH_4, together with small quantities of H_2 and C_2H_6. Traces of C_2H_2 may be formed if a fault is severe or involves electrical contacts. The principal gas is C_2H_4.
2. Overheating of cellulose: Large quantities of CO_2 and CO are evolved from overheated cellulose. Hydrocarbon gases, such as CH_4 and C_2H_4, are formed if a fault involves an oil-immersed structure. The principal gas is CO.
3. Corona: Low-energy electrical discharges produce H_2 and CH_4, with small quantities of C_2H_6 and C_2H_4. Comparable amounts of CO and CO_2 may result from discharges in cellulose. The principal gas is H_2.
4. Arcing: Large amounts of H_2 and C_2H_2 are produced, with minor quantities of CH_4 and C_2H_4. CO_2 and CO may also be formed if a fault involves cellulose. The insulation oil may be carbonised. The principal gas is C_2H_2.

The suggested relationships between key gases and fault types are summarised as follows:

1. O_2 and N_2: non-faults.
2. H_2: corona.

3. CO and CO_2: overheating of cellulose insulation.
4. CH_4 and C_2H_6: low temperature overheating of oil.
5. C_2H_4: high temperature overheating of oil.
6. C_2H_2: arcing.

Since the key gas method does not give numerical correlations between fault types and gas types directly, the diagnosis depends greatly on experience.

6.2.3 Determination of Combustible Gassing Rate

A detected gas volume and its distribution may be generated over a long time period by a relatively insignificant fault or in a very short period of time by a more severe fault. Hence, one measurement does not indicate the rate of gas generation and may provide very little information about the severity of a fault. Once a suspicious gas presence is detected, it is important to sample again and calculate the gassing rate of a gas, which can indicate whether the fault that generated the gas is active or not [3]. The equation for computing gassing rate is as below [5]:

$$R_i = \frac{C_{i2} - C_{i1}}{\Delta t} \times \frac{G}{\rho}, \tag{6.1}$$

where R_i is the gassing rate (ml/h or day), C_{i1} the first sample concentration (ppm), C_{i2} the second sample concentration (ppm), G the total oil weight (ton), ρ the density of oil (ton/m^3) and Δt the actual operating time of a sampling interval (hour or day).

6.2.4 Gas Ratio Methods

As a convenient basis for fault diagnosis, the gas ratio methods are coding schemes that assign certain combinations of codes to specific fault types. The codes are generated by calculating ratios of gas concentrations and comparing the ratios with predefined values, which have been derived from experience and are modified continually. A fault condition is detected when a gas combination fits the code pattern of a particular fault. The most commonly used gas ratio method is the Rogers ratio method [6] as listed in Table 6.1, which is able to distinguish more types of thermal faults than that of the Dörnenberg ratio method [3].

Additional attention should be paid to the following conditions while applying the gas ratio method according to Table 6.1:

1. Significant values quoted for ratio calculations should be only regarded as typical.
2. Transformers fitted with an in-tank OLTC may indicate faults of code 202/102, depending on seepage or transmission of arc decomposition products in the diverter switch tank into the transformer tank oil.

Table 6.1 Fault diagnosis table reproduced from IEC60559

		Code of range of ratios			
		$\frac{C_2H_2}{C_2H_4}$	$\frac{CH_4}{H_2}$	$\frac{C_2H_4}{C_2H_6}$	
	Ratios of gases				
	<0.1	0	1	0	
	0.1–1	1	0	0	
	1–3	1	2	1	
	>3	2	2	2	
Cases	Characteristic fault				Typical examples
0	No fault	0	0	0	Normal aging
1	Partial discharges of low energy density	0	1	0	Discharges in gas-filled cavities resulting from incomplete impregnation, or supersaturation or cavitation or high humidity
2	Partial discharges of high energy density	1	1	0	As above, but leading to tracking or perforation of solid insulation
3	Discharge of low energy	1 → 2	0	1 → 2	Continuous sparking in oil between bad connections of different potential or to floating potential. Breakdown of oil between solid materials
4	Discharge of high energy	1	0	2	Discharges with power follow-through. Arcing - breakdown of oil between windings or coils or between coils to earth. Selector breaking current
5	Thermal fault of low temperature <150°C	0	0	1	General insulated conductor overheating
6	Thermal fault of low temperature range 150°C–300°C	0	2	0	Local overheating of the core due to concentrations of flux. Increasing hot spot temperatures; varying from small hot spots in core, shorting links in core, overheating of copper due to eddy currents, bad contacts/joints (pyrolytic carbon formation) up to core and tank circulating currents
7	Thermal fault of medium temperature range 300°C–700°C	0	2	1	
8	Thermal fault of high temperature >700°C	0	2	2	

3. The ratios for a combination of multiple faults may not fit the predefined codes in Table 6.1.
4. Combinations of ratios not included in Table 6.1 may occur in practice. A great amount of consideration is being given to the interpretation of such combinations.

6.2.5 *Fault Detectability Using Dissolved Gas Analysis*

The majority of faults are slow to develop, which can be detected by DGA monitoring. Locations of faults detectable using DGA as reported by CIGRE [4] are listed below:

1. Within a winding.
2. Cleats and leads.
3. In a tank.
4. A selector switch.
5. A core.

However, instantaneous faults are rapid and sometimes cannot be predicted by DGA. Instantaneous failures that cannot be prevented by DGA are [4]:

1. Flash over with power follow-through.
2. Serious failures, developing within seconds to minutes and therefore not possible to be detected using DGA.

6.3 Combined Criteria for Dissolved Gas Analysis

From a practical point of view, it is important to establish the following evaluation procedures using combined DGA criteria, which are rooted on the guidelines from IEC [2], IEEE [3] and the new modification recommended by CIGRE Task Force 15.01.01 [4].

1. Detection and comparison:
   a. Detect concentrations and gassing rates of any gases dissolved in the oil, and compare them with "normal" quantities using appropriate guidelines. Then it is certain whether an abnormality occurs in a transformer or not.
   b. A recent investigation on DGA undertaken by CIGRE TF 15.01.01 has set up a table with typical values for the key gases H_2, C_2H_2, the sum of the C_1- and C_2-hydrocarbons and the sum of CO_2 and CO for generator and transmission transformers as shown in Table 6.2 [4]. The key gases and the gas concentration values shown in Table 6.2 can be understood as a guideline for DGA interpretation. This is particularly valuable where no additional information is available from historical data of a transformer. However, generally the gassing rate of fault-generated gases is more important than absolute levels, e.g. high levels of key gases can exist with no faults present.
   c. Attention should also be paid to CO and CO_2 when it is suspected that cellulose materials are involved. The high temperature degradation of cellulose, no matter how it is caused (e.g. hot spots or arcing), tends to increase the relative amount of CO. However, the rates of CO_2 and CO production depend greatly upon the oxygen availability, moisture contents and the temperature of degradation. The ratio of $\frac{CO_2}{CO}$ primarily indicates the participation of cellulose insulation materials in electrical or thermal related faults. Normally, in case of overheating of cellulose, the ratio is greater than 10 and in case of degradation of cellulose caused by an electrical fault, the ratio is less than 3 [4].

Table 6.2 Typical values of key gases for generation and transmission transformers

Key gas	Key gas concentration (ppm)	Suspect of indication
C_2H_2	>20	Power discharge
H_2	>100	Partial discharge
$\sum C_xH_y$		Thermal fault
	>1000	if up to $\sum C_1$, C_2, C_3-Hydrocarbons
	>500	if up to $\sum C_1$,C_2-Hydrocarbons
$CO_x = 1,2$	>10000	Cellulose degradation

d. Additionally, a new ratio $\frac{C_2H_2}{H_2}$ is introduced by CIGRE TF 15.01.01 to determine whether fault gases diffuse into a tank from a leaking OLTC or not (if a diverter switch tank and a main tank have a common conservator, it is a similar situation). In this case normally the ratio is 2 and the concentration of C_2H_2 is 30 ppm [4].

e. When an internal fault is suspected, the gas ratio methods and the key gas method should be combined to identify the type of faults, referring to Sects. 6.2.3 and 6.2.4.

2. Assessment: Based upon the combined assessment results using various DGA guidelines, further inspections should be carried out to identify the type and location of faults, such as tests of no-load characteristics of winding DC resistance, insulation, partial discharge and humidity content measurements, followed by the evaluation of the impact of the fault on the serviceability of the observed transformer.
3. Action: Recommended actions should be taken, such as increasing surveillance, shortening sampling intervals, reducing the load on the transformer and finally removing the unit from service.

6.4 Intelligent Diagnostic Methods for Dissolve Gas Analysis

The aforementioned DGA diagnosis techniques reported by IEC, IEEE and CIGRE are computationally straightforward and tend to work well on diagnosing severe faults. They are usually used as a general guideline by experts. However, much uncertainty exists in gas data due to the complexity of gas generating processes in the oil, gas sampling processes and the chromatographic analysis in a laboratory. Consequently, varied patterns and amounts of gases are generated due to different intensities of energy dissipated by different faults, which are affected by oil types, oil temperatures, sampling methods, insulation characteristics and environmental effects, etc. Moreover, most DGA diagnosis techniques rely on experts to interpret DGA results correctly, which could be insensitive to slowly developing and insignificant faults. Sometimes, misjudgement may be made under normal conditions, due to occasional operations, such as oil-tank welding, the electric charge carried by the oil-flow and so forth. To deal with the uncertainties arising from fault diagnosis, based on gas contents extracted from transformer oil samples, various techniques have been attempted by many researchers, including EPS, FL, ANNs and so on. Firstly, fault types are classified based on on-site experience, according to the combined criteria of total combustible gases, gas generation rates, the key gas method and on-site inspections. Subsequently, various CI methods are employed to reveal the relationships between fault symptoms and malfunction types founded on gas-fault mapping schemes.

For instance, ANNs are the most widely used fault classifiers in DGA. In [7, 8], an ANN was utilised to detect faults based merely on previous diagnostic results. Theoretically, an ANN can be trained to represent any observable phenomenon if there are sufficient data available. The experience obtained not only encompasses the existing human diagnostic knowledge, but also explores the unknown relationship between fault conditions and gas data. However, a disadvantage of ANNs in practice is that such relationships are embedded in an ANN structure. It is not easy for a non-expert user to employ these relationships to explain the conclusion of an inference. On the other hand, FL was employed to improve the assessment capability of DGA in [9, 10], which can convert DGA interpretation standards and other human expertise into "if-then" rules to form a decision making system. Moreover, EPSs combined with other CI techniques, e.g. fuzzy models and EAs, have been developed for DGA, which can evaluate ongoing conditions and also suggest proper maintenance actions [11, 12]. The fuzzy set concept can be integrated into an EPS to handle uncertain thresholds, gas ratio boundaries and key gas analysis. EP has also been employed to automatically modify the fuzzy "if–then" rules and simultaneously adjust the corresponding membership functions. Other statistical methods like principal component analysis and correlation analysis were proposed in [13], which are mainly concerned with identifying the key variables and the key gas interdependence of a transformer feature. In summary, these CI techniques could provide a firm heuristic basis for future DGA research.

6.5 Summary

At the beginning of this chapter, a broad literature review of DGA techniques is given, including the theory of fault gas evolution and a variety of conventional gas interpretation schemes. Founded on a literature review of DGA, the correlations between gas types, gas concentrations and fault types are summarised. Then, conventional DGA evaluation procedures are presented briefly and a combined DGA criterion is introduced involving the key gas method, the Rogers ratio method, the gassing rate method and some modifications recommended by CIGRE Task Force 15.01.01. However, based upon the conventional DGA interpretation methods, it is an arduous task to determine malfunction types and oil sampling intervals due to various fault conditions and other interference factors.

Furthermore, establishing relationships between gases and decline conditions is a perplexing task, because of complex gas combination patterns. The DGA procedures and criteria introduced in this chapter are employed in Chaps. 7 and 8 to generate diagnosis inputs for the proposed DGA fault classification systems.

References

1. Heathcote MJ (1998) The J&P transformer book, 12th edn. First published by Johnson & Phillips Ltd, Newnes imprint, UK
2. International Electrotechnical Commission (1978) IEC60559: interpretation of the analysis of gases in transformers and other oil-filled electrical equipment in service. International Electrotechnical Commission Standard, Geneva, Switzerland
3. The Institute of Electrical and Electronics Engineers (1994) Transformers Committee of the IEEE Power Engineering Society, IEEE guide for the interpretation of gases generated in oil immersed transformers, IEEE Std. C57.104–1991. The Institute of Electrical and Electronics Engineers, Inc., New York
4. Mollmann A, Pahlavanpour B (1999) New guidelines for interpretation of dissolved gas analysis in oil-filled transformers. Electra CIGRE France 186:30–51
5. Bureau of Standards for the P.R.China (1987) GB7252-87: Guide for the analysis and diagnosis of gases dissolved in transformer oil, National Technical Committee 44 on Transformer of Standardization Administration of China
6. Rogers RR (1978) IEEE and IEC codes to interpret incipient faults in transformers using gas in oil analysis. IEEE Trans Electr Insul 13(5):348–354
7. Liu YL, Griffin PJ, Zhang Y, Ding X (1996) An artificial neural network approach to transformer fault diagnosis. IEEE Trans Power Deliv 11(4):1838–1841
8. Griffin PJ, Wang ZY, Liu YL (1998) A combined ANN and expert system tool for transformer fault diagnosis. IEEE Trans Power Deliv 13(4):1224–1229
9. Islam SM, Wu T, Ledwich G (2000) A novel fuzzy logic approach to transformer fault diagnosis. IEEE Trans Dielectr Electr Insul 7(2):177–186
10. Yang HT, Liao CC, Chou JH (2001) Fuzzy learning vector quantization networks for power transformer condition assessment. IEEE Trans Dielectr Electr Insul 8(1):143–149
11. Lin CE, Ling JM, Huang CL (1993) An expert system for transformer fault diagnosis and maintenance using dissolved gas analysis. IEEE Trans Power Deliv 8(1):231–238
12. Huang YC, Yang HZ, Huang CL (1997) Developing a new transformer fault diagnosis system through evolutionary fuzzy logic. IEEE Trans Power Deliv 12(2):761–767
13. Mori E et al (1999) Latest diagnostic methods of gas-in-oil analysis for oil-filled transformer in Japan. In: Proceedings of 13th international conference on dielectric liquids (ICDL'99), Nara, Japan, 20–25 July 1999

Chapter 7
Fault Classification for Dissolved Gas Analysis Using Genetic Programming

Abstract This chapter presents an intelligent fault classification approach to transformer DGA for dealing with highly versatile or noise-corrupted data. Two methods, i.e. bootstrap and GP, are employed to preprocess gas data and extract fault features for DGA, respectively. GP is applied to establish classification features for each fault type based on dissolved gases. In order to improve GP performance, bootstrap preprocessing is utilised to equalise the sample numbers for different fault types. The features extracted using GP are then used as inputs fed to ANN, SVM and KNN classifiers for fault classifications. The classification accuracies of integrated GP-ANN, GP-SVM and GP-KNN classifiers are compared with the ones of ANN, SVM and KNN classifiers, respectively. The test results indicate that the developed classifiers using GP and bootstrap can significantly improve diagnosis accuracies for transformer DGA.

7.1 Introduction

As introduced in Chap. 6, DGA has been recognised widely as an effective diagnostic technique to detect incipient faults of power transformers. The analysis of ratios of specific dissolved gas concentrations in the insulation oil of a transformer can indicate the presence of a fault and therefore necessary preventive maintenance arrangements can be scheduled. Currently, there are various DGA interpretation criteria such as the well-known Rogers, modified Rogers, Döernenburg and key gas methods [1]. Using these methods, transformer conditions can be evaluated according to a number of preset thresholds of gases or gas ratios [2]. However, based on empirical studies, these methods are not unbiased and often produce conflicting judgements. In these cases, engineers have to additionally take into account other relevant information about a transformer in

W. H. Tang and Q. H. Wu, *Condition Monitoring and Assessment of Power Transformers Using Computational Intelligence*, Power Systems,
DOI: 10.1007/978-0-85729-052-6_7, © Springer-Verlag London Limited 2011

effort to assess its condition, e.g. the previous operation history of a transformer, results of the latest inspection, states of OLTC and so forth [3].

Apart from the above conventional DGA techniques, various CI techniques have also been investigated extensively for the purpose of developing accurate diagnostic tools using DGA data. Zhang and Ding [4] proposed a two-step ANN method for transformer fault detection using gas concentrations as an input vector to an ANN classifier. Tenfold cross validation [5] of ANN was implemented using 40 sets of samples of gas concentrations. A diagnostic accuracy around 90–95% was achieved by employing an essentially complex ANN structure. The ANN efficiency for the detection of incipient faults in power transformers was analysed by Guardado, et al. [6]. A number of ANN classifiers, trained according to a variety of commonly used practical DGA criteria were compared using a DGA data set of 117 samples for ANN training and 33 new samples for testing. This was reported with a diagnostic accuracy in the range of 87–100% with respect to the criteria used. Huang [7] proposed a transformer assessment technique using a GA-based ANN. A set of 630 real gas ratio records representing five classes was utilised for tenfold cross training and testing [5]. Acquired results showed an accuracy between 90 and 95%.

Moreover, hybrid fault classification methods have been developed for accurate transformer fault diagnosis. A multilevel decision-making model for power transformer fault diagnosis, based on the combination of SVM and *K*NN, was proposed in [8]. An average of 87.5% diagnostic accuracy was obtained by the combined classifier after processing 811 DGA records. In [9] a combination of immune networks and *K*NN was used to process 720 gas samples with a 93.2% diagnostic accuracy. In [10] dissolved gas ratios and its mean, root mean square, variance and higher order central moment values were inputted as diagnostic features to ANN and SVM classifiers using a clonal selection algorithm for feature selections. The analysis of two DGA datasets with 120 and 620 samples showed a diagnostic accuracy in the range of 96–100%. In [11] a combined evolutionary fuzzy diagnosis system processed 561 gas records and delivered a 88% accuracy. A combination of fuzzy logic and ANN was presented in [2] to develop a hybrid transformer fault diagnosis tool. Using the developed classifier more than 80% out of 212 gas samples from 20 transformers were classified correctly. An evidential reasoning approach, based upon the Dempster–Shafer theory and traditional DGA interpretation standards, was utilised for transformer condition assessment taking into account the uncertainties arising from transformer fault diagnoses [12,13].

Regarding artificial feature extractions, GP has been recognised as an efficient technique due to its ability to discover the underlying data relationships and express them mathematically [14]. The ability of GP to generate solutions for complex classification problems has been utilised successfully for machine condition monitoring [15–19] and the combination of GP extracted features with ANN and SVM was reported in [15, 16]. To improve classification accuracies, *K*NN was used as a classifier based on GP extracted features [17]. However, only a few publications of GP applications in the field of transformer fault detection and classification have been reported so far. Zhang and Huang [20] employed GP to develop a binary tree classification structure for DGA sample processing. As a result, an N-class problem was transferred to $N-1$ two class subproblems, where a simple zero threshold was used as a discriminant function to separate the feature space into two regions [20]. The method was trained and tested using 352 gas ratio samples by a tenfold cross validation procedure, which showed a 91.4% accuracy. In [21] the same research team applied GP for DGA data classification to produce discriminant functions and division points. A DGA dataset of 378 samples was used for the validation of the diagnostic method in two formats of input vectors: dissolved gas ratios and actual dissolved gas concentrations being normalised within $[-1, 1]$. The diagnosis delivered 85.4% and 87.8% accuracies, respectively.

Some of the above-reported research were implemented using a large number of gas samples in order to obtain reliable diagnosis performance. Thus, the achieved results could be used for monitoring transformers with the same construction or being operated in similar operation conditions. However, in realistic cases available DGA data contain unequal numbers of samples regarding different fault classes analysed. This can be explained by the lack of data related to different fault conditions. It is possible that collected gas records possess essential diversity even related to the same particular classes due to a large variety of types of transformers investigated and its technical and geometrical characteristics. Furthermore, the noise corruption possibility during measurement procedures is also not easy to avoid. This may lead to lowering of diagnosis accuracies shown in the applications of the above-reported techniques. To overcome the lack of DGA samples, the bootstrap technique was employed as a data preprocessing method prior to GP feature extraction [18, 19].

In this chapter, first the lack of fault class samples is overcome by bootstrap data preprocessing. Then GP is utilised to process gas ratio samples to extract fault features for each class from the available data. The features extracted using GP are then used as the inputs to ANN, SVM and KNN classifiers in order to perform multi-category fault classification. Finally, the comparison between classification accuracies using the proposed approach and that of the individual classifiers, including ANN, SVM and KNN, is given and discussed in detail.

7.2 Bootstrap

Bootstrap was first introduced by Efron as a computer-intensive resampling technique that draws a large number of samples from initial data repeatedly [22]. This data preprocessing technique is designed to obtain reliable standard errors, confidence intervals and other measures of uncertainty in cases when an initial sample number is not sufficient to conduct accurate analysis using other statistical techniques. Because resampling is carried out in a random order, bootstrap assumes no particular distribution of processed data, which gives more applicability with respect to other classical statistical methods [23].

Consider $X^0 = \{x_1,\ldots, x_i,\ldots, x_n\}$, $(i = 1,\ldots, n)$ to be a set of n initial samples with an unknown distribution F, where x_i is the ith independent and identically distributed random variable. Let ϑ denote an unknown characteristic of F, i.e. mean or variance, which is of the interest to be estimated. Then, bootstrap is employed to resample the initial set X^0 in order to obtain k sets of samples $X^j = \{x_{j1},\ldots, x_{ji}, \ldots, x_{jn}\}$, $(j = 1, 2,\ldots, k)$. Each generated sample x_{ji} has a probability n^{-1} of being equally picked up during resampling. Thus, for each generated set of samples X^j, it is possible to calculate an estimator $\widehat{\vartheta}_j$ in order to analyse its probability distribution function $\widehat{F}(\widehat{\vartheta})$ and a confidence interval for the estimator [5, 18, 23]. In this chapter, DGA data are resampled by bootstrap to analyse statistically its diversity and equalise approximately the sample number for each class. These procedures are explained in detail in Sect. 7.4.1. The resampled data are then used for the fault feature extraction and classification using GP and a number of fault classifiers, respectively.

7.3 The Cybernetic Techniques of Computational Intelligence

Cybernetics is a broad field of study and the essential goal of cybernetics is to understand and define functions of various systems. Contemporary cybernetics began in the 1940s as an interdisciplinary study connecting fields of control systems, neuroscience, computer science, mathematics, etc. This section describes briefly the concepts of the ANN, SVM and *K*NN classifiers used for DGA fault classifications, which are among the most practised techniques belonging to the cybernetics techniques of CI.

7.3.1 Artificial Neural Network

Similar to a biological neuron system, ANN is a computational system with a large number of simultaneously functioning simple processes with many connections between nodes of an ANN [24]. ANN renders organisational principles peculiar to a human brain aiming to acquire learning abilities with the purpose of improving its performance. A learning (or training) process for ANN is considered as an iterative adjustment of a network architecture and weights in order to obtain desired outputs for a given set of training samples being passed as ANN inputs. This self-training property makes ANNs more attractive in comparison with other systems which strongly conform to predetermined operational rules formulated by experts.

It has been recognised that one of the most widely used ANN structures for classification problems is multilayer perception (MLP) with a backpropagation learning algorithm [24]. In this chapter, a three-layer MLP structure with input, hidden and output layers is employed as a classifier for transformer fault classification. Each neuron model of the hidden layer has a hyperbolic tangent activation function, whereas a logistic activation function is employed for those of the output layer.

7.3.2 Support Vector Machine

SVM [25] is regarded as one of the standard tools for machine learning and data mining, which is based on advances in statistical learning theories. Originally developed to solve binary classification problems, SVM determines a number of support vectors from training samples and converts them into a feature space using various kernel functions, among which the most commonly used functions are Gaussian radial basis function (RBF), polynomial, multilayer perceptron [25], etc. Thus, for solving a quadratic optimisation problem, the optimal separating hyperplane with a maximal margin between two classes is defined.

For the purpose of multi-category classification, various binary classification methods have been developed, such as "one-against-all", "one-against-one", directed acyclic graph SVM (DAGSVM), etc. [26]. The SVM used in this research is DAGSVM having been approved as one of the appropriate binary methods for multi-category classification [26] with a Gaussian RBF kernel employed and defined by the following equation:

$$K(\mathbf{x}, \mathbf{y}) = \exp\left(-\frac{(\mathbf{x} - \mathbf{y})^2}{2\varsigma^2}\right), \tag{7.1}$$

where $\mathbf{x}$ and $\mathbf{y}$ denote support vectors and ς is an RBF kernel parameter to be predetermined. In order to control the SVM generalisation capability a mis-classification parameter C should also be defined [16].

7.3.3 K-Nearest Neighbour

*K*NN is a supervised learning algorithm that has been used in many applications in the field of data classification, statistical pattern recognition, image processing and many others. It is based upon an assumption of the closest location of observations being members of the same category. The K closest neighbours are found from training datasets by calculating the Euclidean distance between the examined point and training samples. The K closest data points are then analysed to determine which class label is most common among the set in order to assign it to data points being analysed [27]. The value K is selected not to be too small in order to minimise the noise effect in training data. On the other hand, a large value of K essentially increases the computing time; therefore in practice K is adjusted through a number of trials.

7.4 Results and Discussion

In this research, 167 sets of gas samples and matching diagnoses have been extracted from an NG DGA database as the original DGA data, which contain not only seven types of key gases but also diagnosis results from on-site inspections. The transformers being investigated for this study have been evaluated using various engineering diagnostic tools and the corresponding diagnoses related to 4 classes have been generated, i.e. normal unit (NU) (class 1, 26 samples), overheating (OH) (class 2, 69 samples), low-energy discharge (LED) (class 3, 18 samples) and high-energy discharge (HED) (class 4, 54 samples). Based on the DGA analysis technique reported in [6], five commonly used gas ratios are chosen and combined as an input vector for GP feature extractions, which is defined as below:

$$R = \left[\frac{C_2H_2}{C_2H_4}\ \frac{CH_4}{H_2}\ \frac{C_2H_4}{C_2H_6}\ \frac{C_2H_2}{H_2}\ \frac{C_2H_6}{C_2H_2}\right]. \tag{7.2}$$

As a result, each set of individual key gas concentrations is reformatted into a vector in the form of R, and the original DGA data are rearranged as a data array with a dimension of 167×5. Elements of each column of the array are ratios as defined in Eq. (7.2). For instance, the first column of the data array corresponds to the ratio C_2H_2/C_2H_4, being the first element of the input vector R. Then the reformatted DGA data are preprocessed as an initial DGA dataset using bootstrap to derive an expanded DGA dataset containing the four fault classes with equal sample numbers. Subsequently, GP is used to process the expanded DGA dataset for accurate fault feature extractions.

7.4.1 Process DGA Data Using Bootstrap

7.4.1.1 Statistical Analysis for DGA Data

As mentioned above, bootstrap is employed to provide reliable statistical indicators of data in cases where an initial sample number is not sufficient for accurate statistical analysis. In this regard, the initial DGA data are firstly resampled using bootstrap to analyse its diversity. Each column of the reformatted DGA data array, i.e. the initial DGA dataset containing five gas ratios, is considered as a set X^0 of n initial samples ($n = 167$) to generate the expended DGA dataset $\{X^1,\ldots, X^k\}$ consisting of k sample sets.

A publicly available bootstrap toolbox in MATLAB [28] has been used in this research. In Table 7.1 the 95% confidence interval lengths of mean μ and standard deviation σ values for each diagnosis class are presented using bootstrap with the default number of sample sets $k = 199$ as recommended by the bootstrap toolbox [28]. The examination of Table 7.1 shows that there is a large distribution in both μ and σ for all the input vector elements of each class data. Thus, it is

Table 7.1 Length of 95% confidence intervals of gas ratio data

Class	C_2H_2/C_2H_4		CH_4/H_2		C_2H_4/C_2H_6		C_2H_2/H_2		C_2H_6/C_2H_2	
	μ	σ	μ	σ	μ	σ	μ	σ	μ	σ
Normal unit	0.34	0.23	948.37	1.8e4	2.68	3.52	0.52	1.69	9.4e6	1.1e8
Overheating	0.71	1.06	189.49	1021.09	5.78	38.58	3.58	8.00	4.9e5	1.2e6
Low energy discharge	1.35	5.3	1.36	1.02	17.13	80.89	3.45	28.44	2.4e4	1.9e4
High energy discharge	6.06	31.18	0.88	9.55	45.23	1415.37	85.01	3393.07	0.59	8.35

assumed that, based on the available DGA data, it would be sufficiently complex to carry out transformer fault classifications.

7.4.1.2 Sample Number Equalisation for Different Fault Classes

It is logical to expect that more accurate feature extraction with GP can be obtained using a dataset with an equal number of samples regarding each class. In order to approximately equalise the sample number for all the fault classes according to a target sample number, bootstrapping is undertaken for each class by setting a different set number k, which depends on the initial number of the available samples for different classes, and as a result the initial sample number of a class times k equals the target sample number.

Let $n_1,\ldots, n_t, \ldots, n_m$ be the numbers of the initial samples of the corresponding m fault classes in the initial dataset X^0 with a total number of samples n ($\sum_{t=1}^{m} n_t = n$). Set q_{tr} and q_{test} as the target sample numbers for the tth class comprising two datasets, which are generated as a training dataset and a test dataset, respectively. The samples of the tth class are first divided into two different fractions as cn_t and $(1 - c)n_t$, where c represents a fraction coefficient. Then, bootstrapping is performed with the following two set numbers:

$$k_{\text{tr}t} \cong \frac{q_{\text{tr}}}{cn_t} \quad \text{and} \quad k_{\text{test}t} \cong \frac{q_{\text{test}}}{(1-c)n_t}, \tag{7.3}$$

using the corresponding fractions of the tth class to generate two subsets as training and test datasets. The procedure is repeated for all the m classes. As a result, a training dataset and a test dataset with equal sample numbers as q_{tr} and q_{test}, respectively, for each class are derived from the corresponding subsets of each class samples.

Given that, in general, k has to be an integer for data bootstrapping, only integer values, being closest to actual ratios in the right side of the equations in (7.3), are selected. Therefore, the actual numbers of $q_{\text{tr}t}$ and $q_{\text{test}t}$ for each class t may not be equal to the desired values of q_{tr} and q_{test}, but are, in fact, very close to them, respectively.

Following the above procedures, a standard MATLAB function *crossvalind* is selected to divide randomly the initially available gas samples of each class into five data partitions. The four partitions out of five of them are combined to constitute an initial training data fraction and the remaining partition is used as an initial test data fraction. As a result, in total, approximately 80% of the initially available gas data have been used to create a training dataset consisting of 830 samples of gas ratios using bootstrap. The remaining part, approximately 20%, of the initial data is expanded to generate a testing dataset with 228 samples.

In this study, assuming $q_{\mathrm{tr}} = 200$ and $q_{\mathrm{test}} = 50$ as the target values for sample equalisation, the following sample numbers are obtained using bootstrap for the training and test datasets with regard to different classes: normal unit—190 and 70 samples, overheating—200 and 76 samples, low-energy discharge—210 and 42 samples and high-energy discharge—230 and 40 samples, respectively. The slight deviation in the numbers of each class regarding the desired values can be explained by the approximation of $k_{\mathrm{tr}t}$ and $k_{\mathrm{test}t}$ of each class and the utilisation of the MATLAB function *crossvalind* to randomly separate the initial DGA samples into training and testing fractions for different fault classes. As a result, about 78% out of the expanded 1058 samples are employed to constitute a training dataset and 22% of the expanded data are used for testing. It should be noted that the derived training and testing datasets are independent of each other, which are employed to verify the effectiveness of the proposed fault classification approach using GP.

7.4.2 Feature Extraction with Genetic Programming

In order to separate accurately the four fault classes based upon DGA, at least four different fault features are to be extracted by GP. An ECJ software package, i.e a Java-based evolutionary computation software publicly available [29], has been employed to implement GP processing with the parameters listed in Table 7.2. In the following subsections, two GP feature extraction tests are performed using the initial DGA datasets and the expanded datasets with bootstrapping, respectively, in order to verify the effectiveness of data bootstrapping. The extracted features in both the cases are employed to discriminate a fault class from the others. At the end of this subsection a brief discussion is given.

Table 7.2 GP parameters

Parameter	Value	Parameter	Value
No. of generations	100	Crossover probability	0.9
Population size	2,000	Mutation probability	0.1
Maximum depth of tree	10	Number of runs	10
Tournament size	7	No. of elite individuals	1

7.4.2.1 Feature Extraction with the Initial DGA Data

At first, 167 sets of the initial DGA datasets are used for GP feature extractions. A number of GP runs have been carried out, which have shown approximately the same performance. Four of the best GP-generated features to separate each class using the initial DGA data are presented as follows:

$$f_1 = \cos\left(\ln\left(\left|\ln r_2\sqrt{|r_3|}\right|\ln r_1 + r_1 + \sqrt{|r_1|} + (\ln r_3)^2\right)\right), \tag{7.4}$$

$$f_2 = \exp\left(\cos\left(\begin{array}{c}(\sin(\exp(0.3357^{r_1})))^{r_2} + \\ +\sin r_1 + \sqrt{|\sin r_1|} + r_1^{(r_2^2)}\end{array}\right)\bigg/0.5559\right), \tag{7.5}$$

$$f_3 = \sqrt[4]{\left|\sin\left(\left(\sqrt{|r_3|} - (r_3 - r_1)^2\right)^2 \frac{|r_3|r_4^2}{r_1^2}\right)\right|}, \tag{7.6}$$

$$f_4 = \left(\exp\left(\frac{\left\{\begin{array}{c} r_1 \cos r_1 \sin(\sin(\cos r_1)) \\ \times(0.3357 - \sin r_5) + 0.3323 \ln r_2^2 \end{array}\right\}}{-r_1 \ln r_2^2}\right)\right)^{r_5}, \tag{7.7}$$

where $r_1 \sim r_5$ are the five elements of input vector R given by Eq. (7.2), i.e. the dissolved gas ratios.

In Fig. 7.1 the four GP extracted features (7.4)–(7.7) are used to process the initial DGA data in order to discriminate one fault class from the other classes. The first 129 samples belong to the training dataset and the rest 38 ones are employed for validation of the features extracted using GP. All the data have been preliminarily sorted according to its class labels, which are shown in the bottom subfigure of Fig. 7.1. For instance, the DGA data labelled as OH (class 2) are located in the sample range from 21 to 76 in the training datasets and from 136 to 148 in the test datasets. As seen in Fig. 7.1, the majority of samples related to classes 2 and 4, occupying a large portion of the initial DGA data, are well separated from the rest of samples. On the contrary, poor discrimination of samples belonging to classes 1 and 3 can be explained by the lack of samples available for GP training compared with the other two classes.

7.4.2.2 Feature Extraction with the Expanded DGA Data by Bootstrapping

Using the expanded training and testing datasets with approximately equalised number of samples for each fault class, four of the best GP-generated features are obtained:

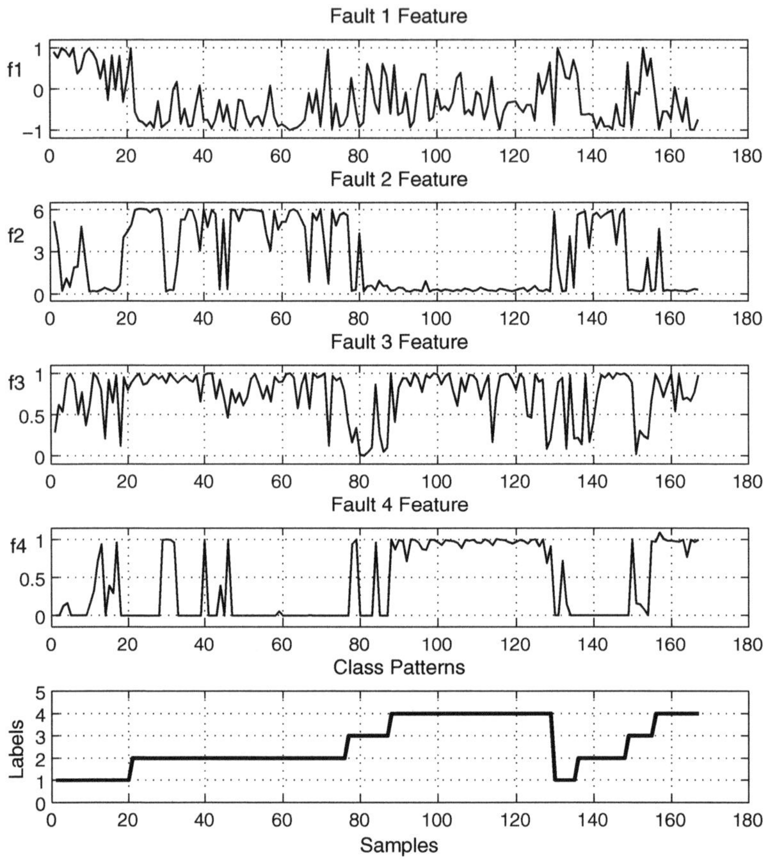

Fig. 7.1 Features extracted by GP (7.4)–(7.7) using the initial DGA data

$$f_{1b} = \arctan\left(\left|\frac{-(r_4r_5)^{r_1^{(r_4r_5)}}}{\arctan(\arctan(\ln r_2))}\right|\right) + \frac{\arctan(r_4r_5)}{\left\{\begin{array}{c}\tan\left(\tan\left(\exp(\sin r_5^{r_1})^{-0.8642\, r_1^{0.1233}}\right)\right) \\ \times \tan\left(\exp(\sin(\cos(\ln(r_4r_5))))^{-\left(r_4r_5r_1^2\right)^{0.1233}}\right)\end{array}\right\}}, \tag{7.8}$$

$$f_{2b} = \sqrt[8]{\left|(\arctan r_1)^{r_2}\right|} - \arctan\left(\left(\arctan\left(\arctan\left(r_2^3\right)^{r_3}\right)\right)^{4r_2^{r_3}}\right), \tag{7.9}$$

$$f_{3b} = -\cos\left(\frac{\cos\left(\frac{\exp r_1}{r_4}\right)}{r_2 r_3 \cos\left(\frac{\cos\left(\frac{\exp r_1}{r_4}\right)}{r_2 r_3}\right)}\right) \Big/ \exp r_1, \tag{7.10}$$

$$\begin{aligned} f_{4b} = {} & \sin(\sin r_1) - \arctan\left(\frac{0.2959\, r_5}{\sin r_1}\right) \\ & - \exp\left(\arctan\left(\frac{0.2959\, r_5}{\sin r_1 - 2r_5}\right)\right) \\ & - \exp\left(\frac{0.0875\, r_5}{\sin r_1 - r_5}\right) - \arctan r_5 \\ & - \exp\left(\frac{r_5 \arctan(\sin(\sin r_1))}{r_1 \sin r_1 - r_1 r_5 - 2r_1 \arctan r_5}\right) \\ & - \arctan\left(\frac{0.2959 \arctan r_5}{\sin\left(\frac{0.0875\, r_5}{\sin r_1 (\sin r_1 - 2r_5)}\right) - r_5}\right). \end{aligned} \tag{7.11}$$

The four GP features (7.8)–(7.11) extracted from the expanded DGA data are given in Fig. 7.2. The first 830 samples belong to the training datasets and the rest 228 ones are used for validation of the features extracted using GP. Class labels are shown in the bottom subfigure of Fig. 7.2. As seen in Fig. 7.2, the majority of samples related to each particular class under consideration are well separated from the rest of samples.

7.4.2.3 Discussions on Feature Extractions with Genetic Programming

Sample overlapping can be observed in both Figs. 7.1 and 7.2, where the samples belonging to one class fall within the numerical region of other classes. This can be explained by the large diversity of the available data of each class. On the other hand, it is presumed that GP features extracted with the expanded DGA data using bootstrap can discriminate the samples belonging to classes 1 and 3 slightly more accurately by comparing corresponding subfigures in Figs. 7.1 and 7.2. This is achieved by approximately equalising the numbers of samples for each fault class for GP training.

In general, the evolvement of GP features largely depends on available training data and the features may not be as effective when processing new data. Therefore, the main purpose of this study is to develop a preprocessing technique that can automatically generate artificial features from DGA samples in order to improve the performance of different fault classifiers.

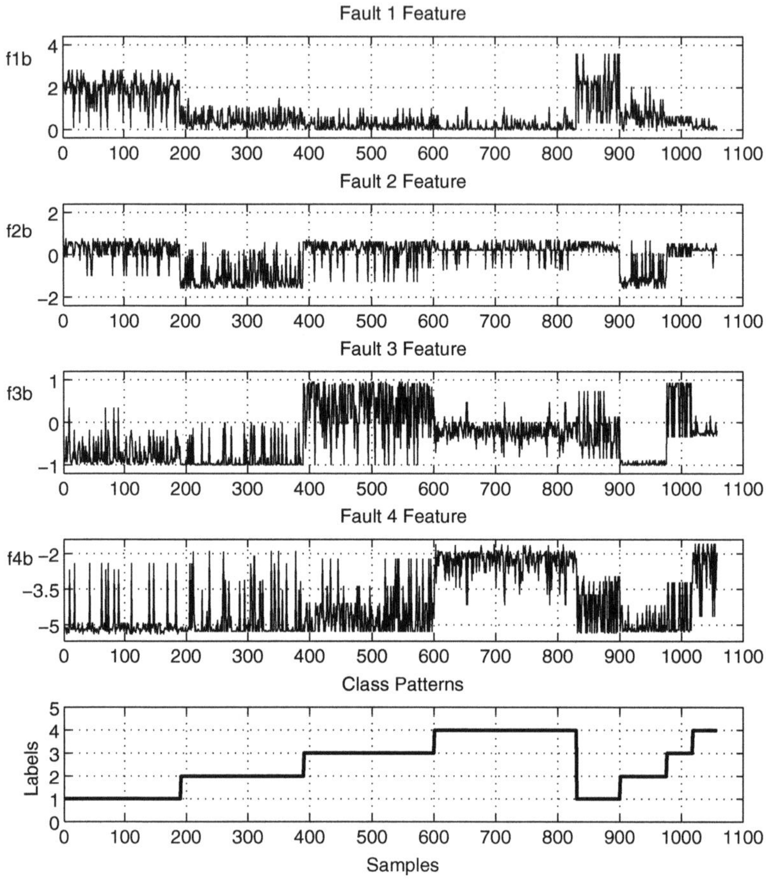

Fig. 7.2 Features extracted by GP (7.8)–(7.11) using the expanded DGA data with bootstrapping

The obtained features (7.4)–(7.7) and (7.8)–(7.11) represent combinations of different gas ratios, thus providing thresholds for more clear separations of one class samples from the others as seen in Figs. 7.1 and 7.2. In this study the features are used to analyse the effectiveness of bootstrap and GP preprocessing for DGA fault classification, which is discussed in the next subsection.

7.4.3 Fault Classification Results and Comparisons

As hybrid classifiers are usually more effective for fault classifications compared with using solely GP features for classification, it is decided to combine GP with the ANN, SVM and *K*NN classifiers to improve the fault classification performance. In this section, the classification results are analysed following the format of the result presentations introduced in [15].

7.4.3.1 Combination of ANN, SVM and KNN with GP without Using Bootstrap

With the purpose to evaluate the classification performance using bootstrap, the GP extracted features (7.4)–(7.7) are employed as inputs to the three classifiers to process the initial DGA data, i.e. an ANN with six neurons in the hidden layer, a SVM with a Gaussian kernel ($\varsigma = 0.0001$ and $C = 2{,}500$) and a *K*NN with a neighbour number $K = 4$. The confusion matrices providing classification accuracies for the four classes are listed in Tables 7.3, 7.4 and 7.5 after processing the initial DGA data. Each row of the tables shows the accuracy percentage with regard to the class samples of the initial DGA data.

From Tables 7.3, 7.4 and 7.5, it is clear that the classification performance is not satisfactory for the NU class with 50.0, 33.3, 50.0% accuracies and the LED class with only 14.3, 28.6, 28.6% for the ANN, SVM and *K*NN classifiers, respectively. In other words, there is a large misclassification rate occurring for the two classes. This is due to the fact that the above two classes have less samples in the initial DGA data with respect to the other classes, which affects GP feature extraction of the corresponding features. Thus, the features (7.4) and (7.6) are not robust enough for sample discrimination regarding different classes. On the other hand, SVM and *K*NN show relatively high classification accuracies for the OH and HED classes with 92.3, 91.7% and 100, 91.7%, respectively. This can be referred to a relatively large number of samples in the initial data available for classifier training. On the whole, unequal numbers of samples for different fault classes for training lead to relatively low overall classification accuracies of 57.89, 71.05 and 76.32% with respect to the three classifiers, using the GP extracted features (7.4)–(7.7) as shown in Tables 7.3, 7.4 and 7.5.

Table 7.3 Test classification accuracy (%) of GP-ANN without using bootstrap

	NU	OH	LED	HED
Normal unit (NU)	**50.0**	33.3	0	16.7
Overheating (OH)	0	**69.2**	30.8	0
Low energy discharge (LED)	14.3	28.5	**14.3**	42.9
High energy discharge (HED)	8.3	8.3	8.3	**75.1**

Table 7.4 Test classification accuracy (%) of GP-SVM without using bootstrap

	NU	OH	LED	HED
Normal unit (NU)	**33.3**	33.3	16.7	16.7
Overheating (OH)	0	**92.3**	7.6	0
Low energy discharge (LED)	42.8	0	**28.6**	28.6
High energy discharge (HED)	0	8.3	0	**91.7**

Table 7.5 Test classification accuracy (%) of GP-*K*NN without using bootstrap

	NU	OH	LED	HED
Normal unit (NU)	**50.0**	16.6	16.7	16.7
Overheating (OH)	0	**100.0**	0	0
Low energy discharge (LED)	42.8	0	**28.6**	28.6
High energy discharge (HED)	0	8.3	0	**91.7**

Table 7.6 Test classification accuracy (%) of GP-ANN using bootstrap

	NU	OH	LED	HED
Normal unit (NU)	**71.4**	14.3	14.3	0
Overheating (OH)	10.5	**86.9**	2.6	0
Low energy discharge (LED)	0	0	**66.7**	33.3
High energy discharge (HED)	0	0	25.0	**75.0**

Table 7.7 Test classification accuracy (%) of GP-SVM using bootstrap

	NU	OH	LED	HED
Normal unit (NU)	**77.1**	0	14.3	8.6
Overheating (OH)	10.5	**89.5**	0	0
Low energy discharge (LED)	0	0	**66.7**	33.3
High energy discharge (HED)	0	0	12.5	**87.5**

7.4.3.2 Combination of ANN, SVM and *K*NN with GP Using Bootstrap

With regard to the expanded DGA data by bootstraping, the GP extracted features (7.8)–(7.11) are processed with the three classifiers, i.e. the ANN and SVM classifiers with the same configurations as indicated in the previous subsection and a *K*NN with $K = 30$. Confusion matrices providing classification accuracies for the four classes are listed in Tables 7.6, 7.7 and 7.8 regarding the expanded DGA datasets.

From Tables 7.6 and 7.7, it is clear that the classification rates of OH samples show a high accuracy for all the classifiers applied. This can be explained as the chemical reactions occurring during an OH fault produce more distinguishable dissolved gas ratios in comparison with the other fault classes. On the other hand, relatively low discrimination rates among LED and HED samples are due to the

Table 7.8 Test classification accuracy (%) of GP-*K*NN using bootstrap

	NU	OH	LED	HED
Normal unit (NU)	**62.9**	0	22.8	14.3
Overheating (OH)	10.5	**89.5**	0	0
Low energy discharge (LED)	0	0	**100.0**	0
High energy discharge (HED)	0	0	25.0	**75.0**

appearance of the same gases generated by the two faults but in slightly different concentrations [1].

Also shown in Tables 7.6, 7.7 and 7.8, ANN, SVM and *K*NN achieve 76.32, 81.14 and 80.7% of total classification accuracies, respectively, by processing bootstrapped DGA datasets with the GP extracted features (7.8)–(7.11). This justifies the effectiveness of bootstrapping for obtaining the approximate equalisation of sample numbers for different classes of the DGA data, which leads to more accurate fault classification rates using GP features.

Considering GP extracted features, despite the better performance of GP extracted features (7.8)–(7.11) in comparison with those being extracted using the initial DGA data, the classification of LED samples shows the lowest accuracy of 66.7% using the ANN and SVM classifiers. In comparison, *K*NN obtains satisfactory sample segregation of the LED class samples and relatively low classification accuracy for the NU data. Therefore, it is reasonable to introduce more GP extracted features for class separation in order to analyse the classification accuracy with different number of GP extracted features applied. Thus, the study in the following section is carried out using only the expanded DGA datasets to improve fault classification accuracies.

7.4.3.3 Effect of Variations in Numbers of GP Extracted Features, ANN Hidden Layer Neurons, SVM Parameters and *K*NN Neighbours

In order to improve DGA fault classification performance, different numbers of GP extracted features have been tried: the four features (7.8)–(7.11), a combination of the four features with two additional GP extracted features for the LED and HED classes (6 features in total), and a combination of features (7.8)–(7.11) with four additional GP extracted features for each class (8 features in total). Furthermore, experiments have been carried out with different classifier structures and its parameters in order to analyse the effect of its variation upon classification performance.

Table 7.9 shows the percentages of fault classification accuracy using the ANN classifier with different numbers of neurons in the hidden layer varying from 3 to 20. For each number of neurons, several experiments have been conducted and an average classification accuracy as 76.32% is obtained with only the four GP extracted features (7.8)–(7.11), whereas the experiment with the eight GP extracted features shows a slightly improved accuracy of 80.26%. The highest accuracy of 85.96% is reached using the least neuron number with the additional two GP extracted features employed. It is clear that too much increase of neuron numbers reduces the classification performance despite of a high training accuracy varied in the range of 92–97%. This refers to overfitting of ANN, when an ANN tends to adapt to the particular details of a specific training dataset [27].

A comparison between Tables 7.9 and 7.10 illustrates that SVM slightly surpasses ANN in terms of classification accuracy with regard to the different number of GP extracted features used. The highest SVM accuracy as 88.16% is also

Table 7.9 Test classification accuracy (%) of GP-ANN with different neuron numbers

	4 GP Features			6 GP Features			8 GP Features			Vector (5)		
	Mean	St.dev	Best	Mean	St.dev	Best	Mean	St.dev	Best	Mean	St.dev	Best
3 Neurons	72.87	1.58	74.60	76.75	7.67	**85.96**	75.09	1.89	77.63	31.32	8.48	40.35
4 Neurons	72.13	2.90	75.44	71.93	2.44	75.44	74.47	6.80	**80.26**	27.72	5.13	33.77
5 Neurons	71.88	2.71	74.75	75.09	2.6	78.51	69.56	2.20	71.49	35.61	8.43	50.0
6 Neurons	69.21	5.22	**76.32**	73.33	5.65	81.14	71.49	4.07	78.07	44.47	7.81	53.95
8 Neurons	57.46	6.52	64.47	64.65	5.98	72.37	63.77	5.88	68.86	40.26	8.22	49.12
10 Neurons	63.25	5.22	68.42	66.41	3.17	70.18	60.88	3.43	63.60	40.96	5.62	50.44
12 Neurons	65.96	6.75	74.12	67.81	3.41	70.18	66.84	7.98	79.39	43.51	8.52	53.95
15 Neurons	60.79	3.97	65.79	63.42	7.95	73.25	66.10	6.47	75.44	42.72	5.93	52.63
20 Neurons	60.08	3.76	64.91	68.77	3.54	74.56	71.32	5.89	79.82	51.67	3.77	**56.58**

Table 7.10 Test classification accuracy (%) of GP-SVM with different SVM parameters

ς	$C = 250$				$C = 2500$			
	GP features			Vector	GP features			Vector
	4	6	8	5	4	6	8	5
0.0001	75.44	79.39	**83.77**	**62.28**	**81.14**	**88.16**	**83.77**	56.14
0.0005	**81.14**	**88.16**	**83.77**	56.14	80.26	81.14	83.33	51.32
0.001	**81.14**	83.77	**83.77**	59.21	76.75	81.14	83.33	51.75
0.005	76.75	85.96	83.33	55.26	73.25	78.95	**83.77**	46.05
0.01	76.75	81.58	**83.77**	55.70	67.11	81.58	80.70	45.18
0.1	73.25	81.58	75.0	60.96	56.14	77.63	75.0	60.96
1	47.81	57.02	62.72	61.84	43.42	57.02	62.72	**61.84**
10	49.12	46.05	47.37	44.30	49.12	46.05	47.37	44.30
100	36.84	35.09	35.53	42.11	36.84	35.09	35.53	42.11

Table 7.11 Test classification accuracy (%) of GP-*K*NN with different neighbour numbers

K	4 GP features	6 GP features	8 GP features	Vector (5)
5	64.47	70.61	74.56	44.30
10	70.61	71.93	83.77	40.79
15	69.74	75.00	81.14	**50.44**
20	68.42	85.53	81.14	48.68
25	80.26	85.09	87.72	45.18
30	**80.70**	82.89	89.91	42.98
35	76.75	87.28	89.91	46.05
40	73.68	89.91	**92.11**	47.37
45	73.68	**92.11**	90.35	46.49
50	73.68	92.11	92.11	46.49
55	73.68	92.11	92.11	42.98
60	73.68	92.11	92.11	42.98

observed with the six GP extracted features using various combinations of SVM parameters.

A *K*NN classifier in combination with GP extracted features delivers the best result compared with the other two classifiers, as is clear from Table 7.11, where the *K*NN classification results are listed with respect to the variations of neighbour number K and the number of GP extracted features used. The utilisation of the six and eight GP extracted features allows *K*NN to surpass ANN and SVM in terms of accuracy and achieves the highest classification accuracy of 92.11% at $K = 45$ and $K = 40$ respectively, which indicates that the combination of *K*NN with GP extracted features is more applicable for this particular task. This can be explained by the fact that the involvement of GP extracted features reduces sample variations within the same classes and increase distances between samples belonging to different classes. This essentially strengthens the performance of a *K*NN classifier, since it classifies data using a similar principle. Consequently, it can be summarised that additional GP extracted features should be applied to discriminate only

Table 7.12 Best classification accuracy (%) of different classifiers

	ANN	SVM	*K*NN
4 GP features	76.32	81.14	80.70
6 GP features	85.96	88.16	**92.11**
8 GP features	80.26	83.77	92.11
Vector (3)	56.58	61.84	50.44

samples of the classes that cannot be clearly separated using a lower number of GP extracted features.

7.4.3.4 Fault Classification Results With and Without GP Extracted Features

Tables 7.9, 7.10 and 7.11 also present the classification results with the three classifiers processing only gas ratios rather than processing GP extracted features. Dissolved gas ratios, given by the input vector defined in Eq. (7.2), have also been processed directly by the ANN, SVM and *K*NN classifiers. The obtained classification results vary in a range of 31–62%, which demonstrate that the individual classifiers are not effective without additional data preprocessing due to essential versatility of the original DGA data. Table 7.12 summarises the highest accuracies derived from the classifiers with respect to different input parameters. Apparently, the *K*NN displays the lowest classification accuracy using the original DGA data without GP processing, whereas it achieves the highest performance in conjunction with six GP extracted features. This is due to the relatively simple *K*NN principle when a classification is undertaken on the basis of the closest distance between examined samples and training dataset with known class labels as explained above in Sect. 7.3.3. Thus, a *K*NN performs well only at consistent data processing with more clear segregation between data categories, which is achieved using the GP extracted features.

7.5 Summary

In this chapter, data preprocessing using bootstrap is employed to overcome the lack of samples of particular fault classes. Then GP is utilised to process gas ratio samples for extracting classification features for each class from the original DGA data. The features extracted using GP are then employed as the inputs to the ANN, SVM and *K*NN classifiers to perform multi-category fault classifications. Comparisons of the classification accuracies between the proposed data preprocessing approach and the ones using individual classifiers without data preprocessing have been given and discussed in detail. The results of the experiments using different numbers of GP extracted features and variations of classifiers' parameters have been discussed, which indicate that the introduction of GP can improve the accuracy of DGA fault diagnosis. The highest accuracy is observed as 92.11% using the GP-*K*NN classifier, which is much better than the one obtained without using GP.

References

1. The Institute of Electrical and Electronics Engineers (1994) Transformers Committee of the IEEE Power Engineering Society, IEEE guide for the interpretation of gases generated in oil immersed transformers, IEEE Std. C57.104-1991. The Institute of Electrical and Electronics Engineers, Inc., 345 East 47th Street, New York, NY 10017, USA
2. Morais DR, Rolim JG (2006) A hybrid tool for detection of incipient faults in transformers based on the dissolved gas analysis of insulating oil. IEEE Trans Power Deliv 21(2):673–680
3. Heathcote MJ (1998) The J&P transformer book, 12th edn. Newnes imprint, UK
4. Zhang Y, Ding X, Liu Y (1996) An artificial neural network approach to transformer fault diagnosis. IEEE Trans Power Deliv 11(4):1836–1841
5. Kohavi R (1995) A study of cross-validation and bootstrap for accuracy estimation and model selection. In: Proceedings of the 14th international joint conference on artificial intelligence (IJCAI), Monreal, Canada, pp 1137–1143
6. Guardado JL, Naredo JL, Moreno P, Fuerte CR (2001) A comparative study of neural network efficiency in power transformers diagnostic using dissolved gas analysis. IEEE Trans Power Deliv 16:643–647
7. Huang YC (2003) Condition assessment of power transformers using genetic-based neural networks. IEE Proc Sci Measure Technol 150(1):19–24
8. Dong M, Xu DK, Li MH, Yan Z (2004) Fault diagnosis model for power transformer based on statistical learning theory and dissolved gas analysis, Conference Record of the 2004 IEEE International Symposium on Electrical Insulation, USA, 85–88
9. Hao X, Sun CX (2007) Artificial immune network classification algorithm for fault diagnosis of power transformer. IEEE Trans Power Deliv 22(2):930–935
10. Cho MY, Lee TF, Kau SW, Shieh CS, Chou CJ (2006) Fault diagnosis of power transformer using SVM/ANN with clonal selection algorithm for features and kernel parameters selection. In: Proceedings of the 1st international conference on innovative computing, information and control (ICICIC'06) 1:26–30
11. Huang YC, Yang HT, Huang CL (1997) Developing a new transformer fault diagnosis system through evolutionary fuzzy logic. IEEE Trans Power Deliv 12(2):761–767
12. Tang WH, Wu QH, Richardson ZJ (2004) An evidential reasoning approach to transformer condition monitoring. IEEE Trans Power Deliv 19(4):1696–1703
13. Spurgeon K, Tang WH, Wu QH, Richardson ZJ, Moss G (2005) Dissolved gas analysis using evidential reasoning. IEE Proc Sci Measure Technol 152(3):110–117
14. Kishore JK, Patnaik LM, Mani V, Agrawal VK (2000) Application of genetic programming for multicategory pattern classification. IEEE Trans Evol Comput 4(3):242–258
15. Guo H, Jack LB, Nandi AK (2005) Feature generation using genetic programming with application to fault classification. IEEE Trans Syst Man Cybern B Cybern 35(1):89–99
16. Zhang L, Jack LB, Nandi AK (2005) Fault detection using genetic programming. Mech Syst Signal Process 19:271–289
17. Zhang L, Jack LB, Nandi AK (2005) Extending genetic programming for multi-class classification by combining k-nearest neighbor. In: Proceedings of the IEEE international conference on acoustics, speech, and signal processing (ICASSP 2005), vol 5, pp V-349–V-352
18. Sun RX, Tsung F, Qu LS (2004) Combining bootstrap and genetic programming for feature discovery in diesel engine diagnosis. Int J Indust Eng 11(3):273–281

19. Li RH, Xie HK, Gao NK, Shi WX (2003) Genetic programming for partial discharge feature construction in large generator diagnosis. In: Proceedings of the 7th international conference on properties of dielectric materials, Nagoya, June, pp 258–261
20. Zhang Z, Huang WH, Xiao DM, Liu YL (2004) Fault detection of power transformers using genetic programming. In: Proceedings of the 3rd international conference on machine learning and cybernetic, Shanghai, China, pp 3018–3022
21. Zhang Z, Xiao DM, Liu YL (2005) Discriminant function for insulation fault diagnosis of power transformers using genetic programming and co-evolution. In: Proceedings of 2005 international symposium on electrical insulation materials, Kitakyushu, pp 881–884
22. Efron B, Tibshirani RJ (1993) An introduction to the bootsrap. Chapman and Hall, New York
23. Zoubir AM, Boashash B (1998) The bootstrap and its application in signal processing. IEEE Signal Process Mag 15:56–76
24. Duda RO, Hart PE, Stork DG (2001) Pattern classification, 2nd edn. Wiley, New York
25. Vapnik VN (1998) Statistical learning theory. Wiley, New York
26. Hsu CW, Lin CJ (2002) A comparison of methods for multiclass support vector machines. IEEE Trans Neural Netw 13(2):415–425
27. Theodoridis S, Koutroumbas K (2003) Pattern recognition, 2nd edn. Academic Press, London
28. Abdelhak MZ, Iskander DR (2007) Bootstrap Matlab Toolbox. http://www.csp.curtin.edu.au/downloads/bootstrap_toolbox.html
29. ECJ 16 (2007) A Java-based evolutionary computation research system. http://cs.gmu.edu/~eclab/projects/ecj/

Chapter 8
Dealing with Uncertainty for Dissolved Gas Analysis

Abstract This chapter presents three approaches to tackling uncertainties occurring in transformer condition assessment, including the ER approach, the hybrid FL and ER approach and the BN approach. Firstly, the methodology of transferring a transformer condition assessment problem into an MADM solution under an ER framework is presented. Three examples for performing transformer condition assessment, using the ER approach, are then illustrated highlighting the potential of the ER approach. The second part of this chapter employs a hybrid approach to the analysis of DGA data based upon several traditional DGA methods. Ideas adapted from the FL theory are applied to soften fault decision boundaries used by the traditional DGA methods. These diagnoses are then considered as pieces of evidence ascertaining to conditions of transformers, which are aggregated using an ER algorithm. The third part is concerned with a BN approach to processing dissolved gases. The methodology of mapping the knowledge in the DGA domain into a BN is described. Finally, an applicable solution to tackle the cases which are not identifiable by the IEEE and IEC code scheme is discussed using the BN approach.

8.1 Introduction

The popularity of DGA stemmed from the fact that DGA tests are conducted without disrupting transformer operations. In short term, high stresses during transformer operations may result in chemical reactions of the oil or cellulose molecules constituting the dielectric insulation, which may be caused by dielectric breakdown of the oil or hot spots. The main degradation products are gases dissolved in the oil which can be detected in the ppm level by on-line or off-line DGA inspections. Besides the DGA technique, other diagnostic techniques are also

W. H. Tang and Q. H. Wu, *Condition Monitoring and Assessment of Power Transformers Using Computational Intelligence*, Power Systems,
DOI: 10.1007/978-0-85729-052-6_8,

employed for detecting transformer faults, e.g. TM, FRA and PDA, which may give different or conflicting fault interpretations for a transformer. Subsequently, how to effectively combine the diagnoses produced from various diagnosis techniques is an interesting issue. It is a similar scenario even for a single DGA task, as different DGA techniques may be involved, e.g. the key gas method, the Rogers ratio method, the gas generating rate method and other industry standards. Most of these diagnostic interpretations are solely done by human experts using experience and standard techniques. Different organisations and companies have their own options and criteria for generating diagnosis reports [1]. Determining the relationships between gas types and faults is a perplexing task, because complex gas combination patterns may arise due to different faults. In order to tackle these complex decision making problems, various CI techniques have been investigated for achieving reliable transformer condition assessment, e.g. ANNs, EPS, EAs, FL and so forth.

Often when dealing with decision making, a decision maker faces uncertainties. In the case of DGA, there are uncertainties arising from vague, imprecise and incomplete diagnoses derived from the traditional DGA methods. As mentioned above, different transformer diagnosis techniques may give different analysis results, and it is difficult for engineers to produce an overall assessment when faced with so much diverse information. Therefore, the combination of available transformer diagnoses to give a balanced overall condition assessment is a very complicated problem and a suitable methodology is required to handle such various diagnostic information. On the other hand, in traditional DGA guidelines crisp decision boundaries are employed for producing fault diagnoses, i.e. the probability of a fault can only be zero or one. In such a manner, the severity and trend of a fault cannot be revealed. Moreover, not all the combinations of gas ratios presented in a fault can be mapped to a fault type in a conventional diagnostic criteria, e.g. the Rogers ratio method. Currently, attentions have been paid to the interpretation of such missing combinations of gas ratios, which are not reported in the relevant IEEE and IEC DGA standards for DGA result interpretations. In this chapter, ER is employed as a decision making framework for integrating results from various DGA diagnostic methods. Fuzzy membership functions are implemented to soften decision boundaries used by gas ratio methods. Finally, a BN is developed for performing probabilistic inference, when dealing with the cases which cannot be identified due to missing codes in conventional DGA diagnosis standards.

8.2 Dissolved Gas Analysis Using Evidential Reasoning

Normally, only one DGA method is selected to provide evaluation results, however, different DGA methods have their own advantages and drawbacks. It is necessary to combine outputs from various DGA diagnostic methods and aggregate these information to form an overall evaluation. If diagnosis results from various DGA criteria are considered as attributes to evaluate transformer conditions, a fault diagnosis process could be regarded as an MADM problem. For such a large quantity of information, how to process it is a complex problem. In addition, as DGA results are sometimes imprecise and even incomplete, a new methodology is required to tackle these uncertainty issues.

8.2.1 A Decision Tree Model under an Evidential Reasoning Framework

8.2.1.1 An MADM Condition Assessment Process

There are various DGA criteria for transformer diagnosis that a power engineer may use to assess transformer conditions by aggregating all available information. For instance, a typical DGA procedure is as follows:

1. Determine TCG presenting in the oil and use the key gas analysis method to check key gas volumes. If the volumes exceed predefined values, a fault is implied.
2. Sometimes neither TCG nor individual gas volumes can indicate the presence of a fault, as a sudden increase in key gases and rate of gas production is more important in evaluating a transformer than the amount of gases. Other complementary DGA techniques have been introduced, e.g. the Rogers ratio method and the gas generating rate method, to detect a transformer fault, which may be able to determine the severity and released energy of a fault.
3. Finally, check recent historical DGA records and decide whether gases are increasing significantly in the observed unit.

It can be seen that a DGA procedure is in fact an evidence combination process, which depends largely on various DGA criteria and engineers' experience. Returning to the problem of transformer condition assessment, a decision tree model can be derived as shown in Fig. 8.1, which is represented graphically in a block diagram form with the root of the tree at the top and leaves at the bottom. The output of the decision tree model in Fig. 8.1 is the overall condition of a transformer. The three elements in the middle level of the tree model represent the thermal condition, the electrical discharge condition and the OLTC condition respectively. The model inputs at the bottom of the tree are derived using different diagnostic techniques, i.e. thermal modelling [2] and various DGA techniques including DGA1 (the key gas analysis method), DGA2 (the CIGRE DGA regulation) and DGA3 (the Rogers ratio method) [1, 3, 4].

Since the mutual dependencies between the branches of the tree model must be considered, it is difficult for a system operator to aggregate all the information shown in the evaluation model of Fig. 8.1. Particularly, it is complex to integrate the outputs of the different levels of the model to form an overall condition evaluation in an analytical manner. Therefore, a supportive decision making tool is

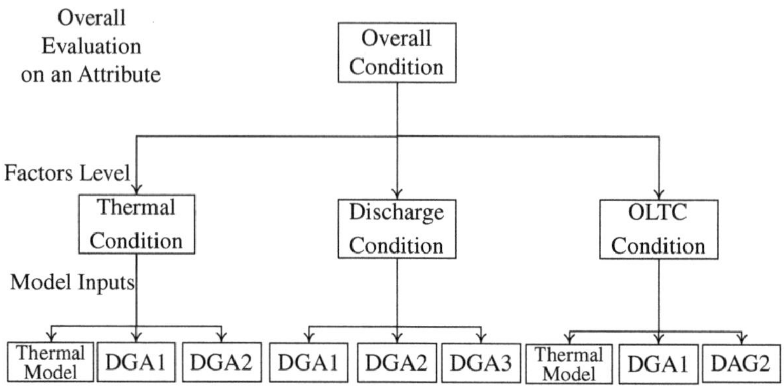

Fig. 8.1 A typical decision tree model for transformer fault evaluation

desirable to deal with such a hierarchical evaluation model. Normally, both quantitative and qualitative information are involved in transformer condition assessment. It is obviously easy to handle quantitative data, compared with qualitative judgements possessing subjective beliefs and uncertainties, hence in this section only qualitative judgements are considered which are the main sources of the uncertainties arising from DGA.

8.2.1.2 Knowledge Transformation into a Decision Tree Model

In order to integrate qualitative judgments effectively and easily, an ER framework is adopted, which can handle uncertainty issues with a firm mathematical foundation. Using ER means that all the subjective data can be organised into an effective and simple decision tree format. For instance, in Fig. 8.1 the root and leaves represent attributes and factors respectively for an MADM problem. The output of the decision tree model is defined as an attribute. The three components in the middle level of the tree model are defined as three composite factors each representing a different facet of transformer conditions:

$$E = \{e_1, e_2, e_3\}. \tag{8.1}$$

At the bottom of the tree, evaluations from a variety of diagnosis methods are listed, i.e. thermal modelling and different DGA methods, whose evaluations are expressed as a set of evaluation grades:

$$H = \{H_1, H_2, H_3, H_4\} = \{\text{Serious, Poor, Normal, Uncertain}\}. \tag{8.2}$$

For instance, considering the "thermal condition" and the "discharge condition" factors, H can be set as {High Temperature (above 400°C), Medium Temperature (above 150°C), Normal (below 150°C), Uncertain} and {High Energy, Low Energy, Normal, Uncertain}, respectively. Using such definitions, the evaluation of each factor of the tree model is expressed in a four-element grade array. Under the ER framework, different diagnosis outputs are weighted accordingly to represent their relative importance during the evaluation, thus a factor or an attribute with the highest weight represents the most important factor or attribute for condition evaluations. In the factor level, each diagnosis may contribute a certain grade to each attribute, then the evaluation grades in different scales can be derived by combining all grade inputs together. Considering the "thermal condition" factor, each diagnosis method under this branch gives a unique evaluation grade, as each grade is distinguishable from the others by setting the boundary values as 400°C and 150°C. For example, if the DGA methods 1, 2 and 3 give independent evaluations as "Medium Temperature", "High Temperature" and "Normal" respectively, assuming without uncertainties, and equal weights are distributed to each method as 1/3, then the inputs in the model input level are given as [0, 0.33, 0, 0], [0.33, 0, 0, 0] and [0, 0, 0.33, 0], respectively, and the overall output from the factor level in this branch is [0.33, 0.33, 0.33, 0]. If a DGA method is only referred to a general fault, e.g. "Thermal Faults", equal probabilities are allocated to both grades as 0.5 on "Medium Temperature" and 0.5 on "High Temperature", respectively, in effect sharing the total belief equally.

If these inputs are considered as evidence for evaluating transformer conditions, then by combining these evidence an overall evaluation or diagnosis of a transformer can be made. In this way, a transformer condition assessment problem is transferred into an MADM solution. An evaluation analysis model based upon ER for diagnosing transformer faults and how to determine the relative weights of factors and attributes are presented in the following sections.

8.2.2 An Evaluation Analysis Model based upon Evidential Reasoning

A hybrid MADM problem for transformer condition assessment may be expressed using the following formula:

$$\text{maximise}_{a\in\Omega} \quad y(a) = [y_1(a), \cdots, y_{k_1}(a), \cdots, y_{k_1+k_2}(a)]. \tag{8.3}$$

where Ω is a discrete set of transformers, ($\Omega = [a_1,\ldots,a_r]$, $r = 1,\ldots,l$), $y(a)$ the overall evaluation of alternative a, $y_k(a)$ the evaluation of the kth attribute of $y(a)$, and k_1 and k_2 the numbers of quantitative and qualitative attributes of each alternative respectively. In this study, alternatives represent a group of transformers and only qualitative judgements are discussed. A decision matrix for qualitative attributes may be presented in Table 8.1 according to Eq. 8.3.

In Table 8.1, y_{rj} are subjective judgements for evaluation of the states of y_j at a_r ($r = 1,\ldots,l$; $j = 1,\ldots,k_2$). The problem is to rank these transformers or to select the best compromise transformer, with the qualitative attributes being satisfied as much as possible.

Table 8.1 An decision matrix for qualitative attributes

Transformers	Qualitative Attributes(y_k)			
(a_r)	y_1	y_2	...	y_{k_2}
a_1	y_{11}	y_{12}	...	y_{1k_2}
a_2	y_{21}	y_{22}	...	y_{2k_2}
...	...	...	...	...
a_l	y_{l1}	y_{l2}	...	y_{lk_2}

In the attribute level of an MADM model, the state of an attribute of each transformer a is required to be evaluated as shown in Fig. 3.1. A simple method for evaluation is to define a few evaluation grades so that the state of an attribute at an alternative can be evaluated to one of the grades. These grades may be quantified using certain scales.

In the evaluation grade level, H_n is called an evaluation grade ($n = 1,...,N$). A set of evaluation grades specified for an attribute y_k is denoted by

$$H = \{H_1, \ldots, H_n, \ldots, H_N\}, \tag{8.4}$$

where N is the number of evaluation grades. In this study, Eq. 8.4 is a mapping of Eq. 8.2.

In the factor level, E_k represents a set of factors which is associated with the evaluation of the basic attribute $y_k(a)$ and denoted by

$$E = \left\{e_k^1, e_k^2, \ldots, e_k^{L_k}\right\}, \quad k = 1, \ldots, k_2, \tag{8.5}$$

where e_k^i ($i = 1,...,L_k$) are factors influencing the evaluation of $y_k(a)$. The state of e_k^i can be directly evaluated at an alternative a, that is $e_k^i = e_k^i(a)$. A larger preference degree value is interpreted as a higher evaluation grade. So, the preference degree for the state of an attribute $y_k(a)$ through the direct evaluations of the relevant factors e_k^i can then be generated and integrated by using the Dempster-Shafer theory presented below.

As stated in Sect. 3.2.1.2, under an ER framework the overall probability assignment can be derived by combining all the basic probability assignments using the operational algorithms below. Define a factor subset $e_{I_k(i)}(a)$ and a combined probability assignment $m^{\Psi}_{I_k(i)}(a)$ as follows:

$$\begin{aligned} e_{I_k(i)}(a) &= \{e_k^1(a), \ldots, e_k^i(a)\}, \ 1 \le i \le L_k, \\ m^{\Psi}_{I_k(i)}(a) &= m(\Psi/e_{I_k(i)}(a)) = m_i^{\Psi}(a), \end{aligned} \tag{8.6}$$

where $m(\Psi/e_{I_k(i)}(a))$ is the combined probability assignment to Ψ confirmed by $e_{I_k(i)}(a)$.

To combine $e_{I_k(2)}(a) = \{e_k^1(a), e_k^2(a)\}$, an intersection tableau is constructed in Table 8.2 as an example. From the combination rules defined in Chap. 3 , we can obtain the following recursive formulae:

Table 8.2 Intersection tableau of combining $\{e_k^1(a), e_k^2(a)\}$

$e_{I_k(2)}$		e_k^2			
		$\{H_1\}(m_{k2}^1)$	$\cdots$	$\{H_n\}(m_{k2}^n)$	$\{H\}(m_{k2}^H)$
e_k^1	$\{H_1\}(m_{k1}^1)$	$\{H_1\}(m_{k1}^1 m_{k2}^1)$	$\cdots$	$\{\Phi\}(m_{k1}^1 m_{k2}^n)$	$\{H_1\}(m_{k1}^1 m_{k2}^H)$
	$\cdots$	$\cdots$	$\cdots$	$\cdots$	$\cdots$
	$\{H_n\}(m_{k1}^n)$	$\{\Phi\}(m_{k1}^n m_{k2}^1)$	$\cdots$	$\{H_n\}(m_{k1}^n m_{k2}^n)$	$\{H_n\}(m_{k1}^n m_{k2}^H)$
	$\cdots$	$\cdots$	$\cdots$	$\cdots$	$\cdots$
	$\{H_N\}(m_{k1}^N)$	$\{\Phi\}(m_{k1}^N m_{k2}^1)$	$\cdots$	$\{\Phi\}(m_{k1}^N m_{k2}^n)$	$\{H_N\}(m_{k1}^N m_{k2}^H)$
	$\{H\}(m_{k1}^H)$	$\{H_1\}(m_{k1}^H m_{k2}^1)$	$\cdots$	$\{H_n\}(m_{k1}^H m_{k2}^n)$	$\{H\}(m_{k1}^H m_{k2}^H)$

$$\{H_n\} : m_{I_k(i+1)}^n = K_{I_k(i+1)}\left(m_{I_k(i)}^n m_{k,i+1}^n + m_{I_k(i)}^n m_{k,i+1}^H + m_{I_k(i)}^H m_{k,i+1}^n\right), n = 1, \ldots, N, \tag{8.7}$$

$$\{H\} : m_{I_k(i+1)}^H = K_{I_k(i+1)} m_{I_k(i)}^H m_{k,i+1}^H, \tag{8.8}$$

where

$$K_{I_k(i+1)} = \left[1 - \sum_{\tau=1}^{N} \sum_{j=1, j\neq\tau}^{N} m_{I_k(i)}^{\tau} m_{k,i+1}^{j}\right]^{-1}, \quad i = 1, \ldots, L_k - 1. \tag{8.9}$$

Obviously, $m_{I_k(i+1)}^{\Psi} = 0$ for any $\Psi \subseteq H$ other than $\Psi = H_n$ ($n = 1,\ldots,N$) and H. It can be proven from the combination procedure that $m_{I_k(L_k)}^{\Psi}$ is the overall probability assignment to $\Psi(\subseteq H)$ confirmed by $E_k(a)$ and $m_{I_k(L_k)}^{\Psi} = 0$ for any $\Psi \subseteq H$ other than $\Psi = H_n$ ($n = 1,\ldots,N$) and H. Consequently, the overall preference degree of alternative a_r may be calculated using the following equation:

$$p_{rk} = p(y_k(a_r)) = \sum_{n=1}^{N} m_{I_k(L_k)}^n p(H_n) + m_{I_k(L_k)}^H p(H), \tag{8.10}$$

where

$$p(H) = \sum_{n=1}^{N} p(H_n)/N.$$

A larger preference degree maps to a higher evaluation grade according to Eq. 3.6.

8.2.3 Determination of Weights of Attributes and Factors

In order to implement the original ER combination algorithm, a set of weights to reflect the relative importance of each basic attribute should be defined. In general, the weights for basic attributes can be assigned directly by experts based on

Table 8.3 Scale of measurement for AHP

Values	Definition
1	Equally important or preferred
3	Slightly more important or preferred
5	Strongly more important or preferred
7	Very strongly more important or preferred
9	Extremely more important or preferred
2, 4, 6, 8	Intermediate values to reflect compromise
Reciprocals	Used to reflect dominance of the second alternative as compared with the first

experience or be derived using a method of pairwise comparison of attributes such as the eigenvector method [5]. In this study, paired comparison judgements used in an analytic hierarchy process (AHP) are applied to pairs of homogeneous elements concerning subjective criteria. According to AHP, attributes are compared in a pairwise manner using a predefined fundamental scale of values to represent intensities of judgements, which are presented in Table 8.3 [5]. It is necessary to ensure that the criteria are all on the same scale, otherwise the weighted importance of the criteria would be meaningless.

In order to obtain a vector of priorities or relative weights of attributes, a pairwise comparison matrix **A** is defined, where each element $a_{i,j}$ $(i, j = 1,...,L)$ represents the relative importance of attribute i over attribute j using the scale defined in Table 8.3. A general comparison matrix is shown as:

$$\mathbf{A} = \begin{bmatrix} a_{1,1} & a_{1,2} & \cdots & a_{1,L} \\ a_{2,1} & a_{2,2} & \cdots & a_{2,L} \\ \vdots & \vdots & \ddots & \vdots \\ a_{L,1} & a_{L,2} & \cdots & a_{L,L} \end{bmatrix} \tag{8.11}$$

Normalisation of **A** is performed via dividing each element $a_{i,j}$ of the matrix by the sum of all the elements in the corresponding column j. As a result, a normalised matrix $\bar{\mathbf{A}}$ is derived, and finally the arithmetic mean value of each row i of the normalised matrix $\bar{\mathbf{A}}$ represents the relative weight ω_i of the corresponding attribute e_i $(i = 1,...,L)$.

Using the matrix of ratio comparisons, the vector of priorities can then be derived by the eigenvector method. Subsequently, the weights are the eigenvector solution of its principal eigenvector value. For example, suppose that in order to determine the overall condition of a transformer three factors are used involving the thermal condition, the PD condition and the OLTC condition. On the basis of Table 8.3, a comparison matrix can be constructed as below:

$$\mathbf{A}_{mn} = \begin{bmatrix} 1/1 & 1/2 & 3/1 \\ 2/1 & 1/1 & 4/1 \\ 1/3 & 1/4 & 1/1 \end{bmatrix}. \tag{8.12}$$

Each pair-comparison element in $\mathbf{A}_{mn}$ $(m, n = 1,\ldots,L)$ represents the relative importance of factor m over factor n. Then the principal eigenvector solution of the above matrix is [0.489, 0.848, 0.203], which is the priority vector λ_k $(k = 1, 2, 3)$ of the three factors as illustrated in Fig. 8.1.

8.2.4 Evaluation Examples under an Evidential Reasoning Framework

In this section, the ER approach is utilised to assess the condition of a power transformer, as well as to rank conditions of a group of transformers for power system maintenance purposes. As discussed in Sect. 8.2.1.1, the overall condition of a transformer can be assessed by considering three aspects, i.e. the thermal condition, the discharge condition and the OLTC condition. If the fault diagnoses of the three aspects can be obtained separately, they may then be treated as pieces of evidence, allowing the overall condition of the transformer to be assessed by aggregating these evidence. Based upon this, the ER approach is employed to produce evaluations for power transformers with a decision tree model as shown in Fig. 8.1. Several practical applications are now illustrated and discussed as the following.

8.2.4.1 A Simple Example for Assessing Conditions of the Same Unit at Different Time Stamps

Firstly, a simple example using the ER approach is demonstrated to evaluate conditions of a power transformer at different time stamps. Several sets of DGA data were sampled from an on-site transformer, SGT4 (240MVA, 400/132 kV), in 09/1997 and 10/1997, respectively, which are listed in Table 8.4, where P1 represents the sample point in the main tank and P2 the sample point in

Table 8.4 DGA concentrations of a scrapped transformer (SGT4)

ppm	P1(09/1997)	P2(09/1997)	P1(10/1997)	P2(10/1997)
CO	474	469	522	553
CH_4	29.0	17.0	78.0	210
CO_2	2,761	2,739	3,044	3,069
C_2H_4	14.0	10.0	47.0	109.0
C_2H_6	11.0	7.0	21.0	54.0
C_2H_2	0.2	0.7	49.8	112.9
H_2	22.0	28.0	132	329
O_2	5,394	3,695	13,980	12,737
N_2	51,001	58,947	57,240	58,315
H_2O	15.0	25.0	13.0	13.0

the selector. The tap changer of SGT4 exploded on 07/10/1997 and the tap winding on the C phase was heavily distorted due to the failure of the tap changer.

The ER process for evaluating the condition of a transformer is presented as follows. Firstly, subjective judgements are produced using various DGA methods for all the alternatives involved. The weight for each attribute is determined and factored using the AHP method or expertise provided by on-site engineers. The subjective judgements are scaled into weighted outputs under an ER framework, and the Dempster-Shafer combination rules are applied to derive an overall evaluation, which is predefined as a set of grades. Finally, the preference degrees for each transformer are calculated and used to rank these alternatives' conditions. The weights of the three factors have been derived in Sect. 8.2.3, noticing that only qualitative factors are involved in this example. Three types of DGA methods are chosen, i.e. the key gas method, the CIGRE DGA regulation and the Rogers ratio method. The outputs of the three DGA methods are fed into the decision-tree model as evidence for evaluating the thermal condition, the OLTC condition and the discharge condition. As no temperature sensors were fitted on SGT4, the model inputs from thermal modelling are set as zero, which means that the diagnostic information is incomplete in decision making.

For the oil in the main tank sampled in 09/1997, where the inputs from the thermal model are not available, the CIGRE DGA regulation diagnoses as "thermal faults" due to "$C_2H_2/C_2H_6 > 1$", while the Rogers ratio method generates a code [1,2,1] interpreted as "medium thermal faults". Thus, for the "thermal condition" branch, the inputs of three bottom "leaves" are [0, 0, 0], [0.5, 0.5, 0] and [0, 1, 0], respectively. As each "leaf" is given the same weight 1/3, the overall input of this branch is $[(0 + 0.5 + 0)/3, (0 + 0.5 + 1)/3, (0 + 0 + 0)/3] = [0.167, 0.5, 0]$. For the "discharge condition" branch, all the three DGA methods report as "normal condition", thus the input of this branch is [0, 0, 1.0]. Similarly, for the oil from the selector sampled in 09/1997, no inputs from "thermal model" are provided, the CIGRE DGA regulation diagnoses "thermal faults" due to "$C_2H_4/C_2H_6 > 1$", and the Rogers ratio method gives a code [0, 1, 1] with no match to its interpretation table. Therefore, for the "OLTC condition" branch, the inputs of three "leaves" are [0, 0, 0], [0.5, 0.5, 0] and [0, 0, 0], respectively. As each "leaf" is given the same weight 1/3, the overall input of this branch is [0.167, 0.167, 0].

For the oil from the main tank sampled in 10/1997, the CIGRE DGA regulation also diagnoses as "discharge fault" and "thermal fault" due to "$C_2H_2/C_2H_6 > 1$" and "$C_2H_4/C_2H_6 > 1$", and the Rogers ratio method gives a code [1, 0, 1] interpreted as "discharge of low energy". The key gas method reports a fault as "discharge faults", because both the concentrations of C_2H_2 and H_2 exceed the guideline limits. Thus, for the "thermal condition" branch, the inputs of the three bottom "leaves" are [0, 0, 0], [0, 0, 1.0] and [0.5, 0.5, 0] respectively. As each "leaf" is given the same weight 1/3, the overall input of this branch is [0.167, 0.167, 0.333]. For the "discharge condition" branch, the inputs of "leaves" are [0.5, 0.5, 0], [0, 1.0, 0] and [0.5, 0.5, 0] respectively, so the overall input of this branch is [0.333, 0.50, 0]. Similarly, for the "OLTC condition" branch, the same results are derived, and the overall input is [0.333, 0.50, 0].

Table 8.5 The decision matrix of a simple evaluation example

Confidence degrees		SGT4 09/1997			SGT4 10/1997		
		H_1	H_2	H_3	H_1	H_2	H_3
Factors	e_1	0.17	0.5	0.0	0.17	0.17	0.33
	e_2	0.0	0.0	1.0	0.33	0.5	0.0
	e_3	0.17	0.17	0.0	0.33	0.5	0.0

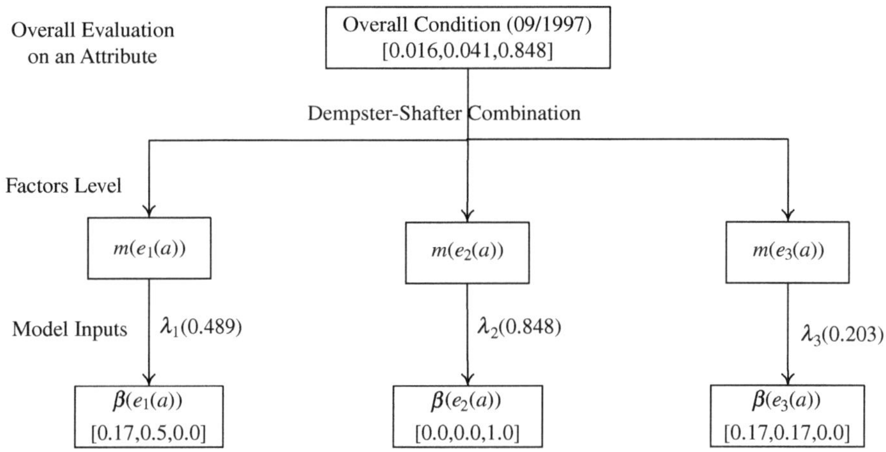

Fig. 8.2 Evidence combination processes for DGA data in 09/1997

The decision matrix under the ER framework is shown in Table 8.5 by reorganising the derived diagnosis outputs.

In order to interpret these outputs into scaled inputs, the priority vector produced in Sect. 8.2.3 is assigned to the factors in each branch. After ER calculations, the overall evaluations of SGT4 at different periods of time are: for the sample in 09/1997 it is [0.016, 0.041, 0.848, 0.01], whose evidence combination processes are shown in Fig. 8.2; and for the sample in 10/1997 it is [0.304, 0.462, 0.048, 0.186]. Comparing these results under the ER framework, these four elements in the outputs are in accordance with the grade set of [Serious, Poor, Normal, Uncertain], representing the condition of a transformer. Considering the grade scales, it is apparent that the condition in 09/1997 is better than the one in 10/1997, as high fault probabilities are indicated for the sample in 10/1997, which matches the actual on-site investigation.

8.2.4.2 A Generic Example to Evaluate Conditions for Different Transformers

A generic evaluation example concerning different transformers, i.e. $\Omega = [a_1, a_2, a_3]$, is illustrated in this section. In order to simplify the analysis

Table 8.6 The decision matrix of a generic evaluation example

Confidence degrees		Transformer a_1			Transformer a_2			Transformer a_3		
		H_1	H_2	H_3	H_1	H_2	H_3	H_1	H_2	H_3
Factors	e_1	0.7	0.2	0.1	0.0	0.2	0.8	0.7	0.3	0.0
	e_2	0.0	0.8	0.1	0.0	0.4	0.6	0.5	0.5	0.0
	e_3	0.0	0.0	0.8	0.0	0.0	0.7	0.2	0.7	0.0

processes, only the inputs at the bottom of the ER decision tree are given, and the subjective judgements for assessments are shown in Table 8.6 for an illustration purpose. It is noticed that the provided information is incomplete and therefore contains uncertainties, which is derived from different diagnostic methods. After ER calculations using the same weight scales as derived in Sect. 8.2.3, the overall evaluations of each transformer are: [0.098, 0.644, 0.123, 0.135] for a_1, [0.0, 0.252, 0.693, 0.055] for a_2 and [0.517, 0.425, 0.0, 0.058] for a_3. By using Eq. 8.10, the preference degree for each transformer can be derived, e.g. the preference degree of a_1 is calculated as $p(y_k(a_1)) = 0.098 \times (-1) + 0.644 \times 0 + 0.123 \times 1 + 0.135 \times 0 = 0.025$ by setting $p[H_n] = [-1, 0, 1]$. Finally, the preference degrees of each transformer are [0.025, 0.693, −0.517], i.e. $[p(y_k(a_2)) > p(y_k(a_1)) > p(y_k(a_3))]$, which means that transformer a_2 is in the best condition among the three transformers.

This generic evaluation example demonstrates that the ER approach is able to treat uncertain decision knowledge in a clear and simple manner. The employed evidence combination algorithms are useful for combining multiple uncertain subjective judgements.

8.2.4.3 A More Complex Condition Assessment Example

For a real transformer assessment problem, the above two examples are not applicable, as more factors may be involved in a real situation. If more attributes and factors are involved, the ER approach is still capable of tackling these problems in a meaningful and simple manner. In Fig. 8.3, a more complex example for diagnosing transformer conditions is shown.

There are now three attributes in the decision tree and more factors are involved. The overall condition of the observed unit may then be determined on the basis of the following three components, i.e. the oil condition, the winding condition, the OLTC condition, which are in the attributes level. In the factor level, the inputs of the tree model are again derived from a variety of transformer diagnostic methods. The uncertain subjective judgements are presented in Table 8.7, where y_k ($k = 1, 2, 3$) are the attributes and e_k^i ($k, i = 1, 2, 3$) are the factors involved. The weights of each factor are chosen as equal for an illustration purpose, e.g. 1/3. Alternatively, the weights could be determined by engineers based on their on-site experience using the AHP method.

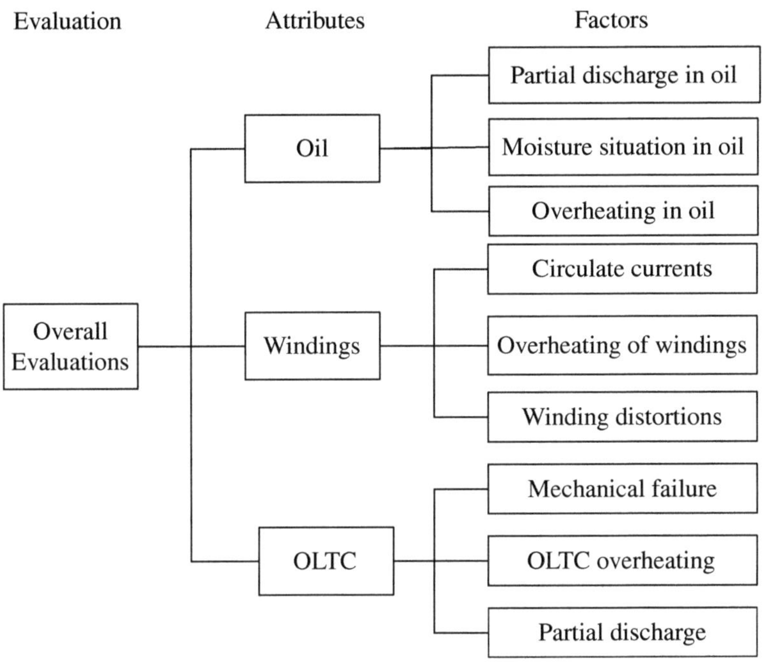

Fig. 8.3 A complex evaluation example using an expanded ER decision tree-model

Table 8.7 The subjective judgments for condition assessments of a complex evaluation example

Attributes level	Factors level	Unit condition		
		H_1	H_2	H_3
Oil (y_1)	e_1^1	0.2	0.7	0.0
	e_1^2	0.1	0.9	0.0
	e_1^3	0.3	0.7	0.0
Winding (y_2)	e_2^1	0.0	0.0	1.0
	e_2^2	0.3	0.7	0.0
	e_2^3	0.1	0.8	0.1
OLTC (y_3)	e_3^1	0.0	1.0	0.0
	e_3^2	0.8	0.1	0.0
	e_3^3	0.5	0.5	0.0

After ER calculations using the data in Table 8.7, the overall assessment matrix of the observed transformer regarding each evaluation attribute is:

$$\begin{bmatrix} & H_1 & H_2 & H_3 & H_4 \\ y_1(a) & 0.042 & 0.954 & 0.0 & 0.003 \\ y_2(a) & 0.072 & 0.702 & 0.215 & 0.012 \\ y_3(a) & 0.238 & 0.753 & 0.0 & 0.009 \end{bmatrix} \tag{8.13}$$

Comparing these results with the ER grade definition, the four elements in each row of the output matrix are mapped to a grade set of [Serious, Poor, Normal, Uncertain] for each attribute, which represent the condition of each evaluation attribute. The preference degree of each attribute is generated as [−0.042, 0.143, −0.238], which is quite close to the grade value of the poor condition definition and indicates the observed unit is very likely to be in a poor condition. Based on this assessment, cautions should be taken and an outage may be planned accordingly. With regard to the number of the factors involved, sometimes even the information in the bottom leave level is not available in some branches of the decision tree. For such an MADM problem the ER approach can still handle it properly by adjusting the decision tree structure, which further demonstrates its capability in the applications of transformer condition assessment. The results also demonstrate the properties of the Dempster–Shafer theory and its potential to treat uncertainties in MADM problems through multiple factor analysis using ER.

8.3 A Hybrid Diagnostic Approach Combining Fuzzy Logic and Evidential Reasoning

As stated previously, when DGA reviewers examine dissolved gases, they compare the values that they have with the decision rules of several traditional analysis methods. What an engineer then does is to make subjective decisions and allowances, i.e. does the data fit any of the decision criteria and if not how close is the data to those criteria? The proposed hybrid diagnostic approach detailed in this section can generate subjective judgements by using fuzzy membership functions to soften the decision boundaries which are currently utilised by the traditional DGA methods. More precisely, crisp decision boundaries imply that the probability of a fault can only be zero or one, i.e. $\Pr(\text{Fault}) \in \{0,1\}$. Softening such boundaries using appropriate functions means that the probability of a fault can take on any value in a closed interval [0, 1]. By setting the boundaries in this way, the belief that an engineer has that a transformer is faulty can be represented by a single value, $\Pr(\text{Fault}) \in [0,1]$.

The boundary conversions implemented in this study are largely intuitive and based on summaries drawn from inspecting the Halstead's thermal equilibrium diagram shown in Fig. 8.4 taken from the IEEE American Standard C57.104-1991 [4]. These fuzzy sets are able to generate a set of possible faults and associated probabilities, as opposed to an oversimplified set of results generated by the traditional DGA methods, i.e. {Fault, No Fault, No Decision}. These new probabilistic results are then treated as pieces of evidence ascertaining to transformer conditions and combined using the original ER algorithm as discussed in Chap. 3 to produce an overall evaluation of transformer conditions.

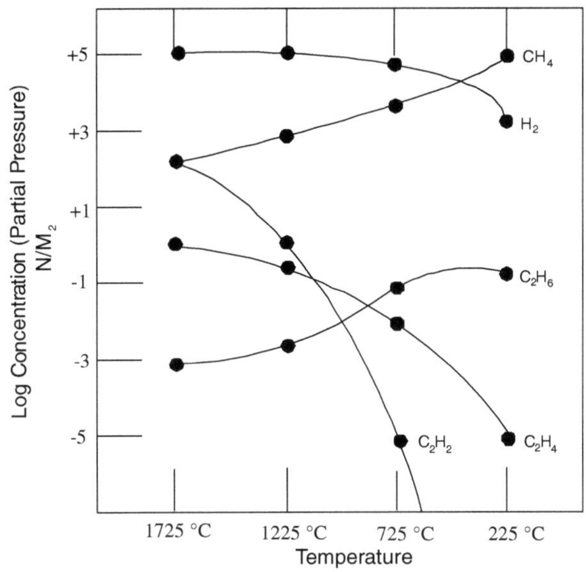

Fig. 8.4 Halstead's thermal equilibrium of partial pressures [4]

8.3.1 Solution to Crispy Decision Boundaries

The three traditional DGA methods incorporated in this study are the Rogers ratio method (RRM), the Dörnenburg's ratio method (DRM) and the key gas method (KGM). All the three methods have their theory routed in organic chemistry and base their diagnoses on matching the temperature generated by a fault to a general fault type. Put simply, each fault type typically generates a fault temperature within a prescribed range, and the more severe the fault the higher the temperature. Because the insulation oil used in power transformers is organic (i.e. composed primarily of hydrocarbons), certain fingerprint gases are generated at specific temperature ranges, allowing the traditional methods to identify a possible fault temperature range and a possible fault type. KGM actually employs four characteristic charts which represent typical relative gas concentrations for four general fault types: overheating of cellulose (OHC), overheating of oil (OHO), partial discharge (PD) or arcing. The other two methods use ratios of fingerprint gases to try and pinpoint specific temperature ranges. The fingerprint gases used are CO, H_2, CH_4, C_2H_6, C_2H_4 and C_2H_2. Figure 8.4 shows the Halstead's thermal equilibrium chart, upon which the theory for the three methods is based. What the diagram illustrates is the theoretical equilibrium of partial pressures of the fingerprint gases excluding CO. It must be noted that when dealing with gaseous chemicals, the relative partial pressures of gases are equivalent to the relative concentrations of the gases. So, from this chart key ratios have been devised which are used to identify certain fault types. These ratios are: $R_1 = CH_4/H_2$, $R_2 = C_2H_2/C_2H_4$, $R_3 = C_2H_2/CH_4$, $R_4 = C_2H_6/C_2H_2$ and $R_5 = C_2H_4/C_2H_6$. Tables 8.8 and 8.9 show the decision rules used by RRM and DRM respectively. For the DRM diagnosis to be credible

Table 8.8 Rogers ratio method [6]

Fault	R_1	R_2	R_5
No fault	0.1–1.0	<0.1	<1.0
PD	<0.1	<0.1	<1.0
Arcing	0.1–1.0	0.1–3.0	>3.0
Low temp. thermal	0.1–1.0	<0.1	1.0–3.0
Thermal <700°C	0.1–1.0	<0.1	1.0–3.0
Thermal >700°C	>1.0	<0.1	>3.0

Table 8.9 Dörnenburg's ratio method [4]

Fault	R_1	R_2	R_3	R_4
Thermal	>1.0	<0.75	<0.3	>0.4
PD	<0.1	N/A	<0.3	>0.4
Arcing	>0.1 and <1.0	>0.75	>0.3	<0.4

Table 8.10 Dörnenburg's L1 limits [4]

Gas	H_2	CH_4	CO	C_2H_2	C_2H_4	C_2H_6
L1 (ppm)	100	120	350	35	50	65

the levels of the gases used in the ratios must be greater than some predetermined levels (in ppm) known as the L1 limits, which are listed in Table 8.10. Figure 8.5 shows the four general fault scenarios used in KGM.

Both of the two ratio methods make decisions based on the crisp values of the ratios that they use. The problem this creates is best illustrated by the following simple example. Consider the case when using RRM, R_2 and R_5 are within the prescribed limits of a normal operating condition but $R_1 = 1.1$. The decision rules state that the value of R_1 is outside the normal operating condition and therefore the No Decision result is returned since these values do not match any of the other fault conditions. However, a further inspection of the values shows that the value of R_1 is in fact very close to the decision boundary and that the transformer is more likely to be in a normal operating condition. This thought process can be mimicked by softening the decision boundaries set out in Tables 8.8 and 8.9, by converting all of the crisp boundaries into fuzzy set equivalents. The result is that the values of R_2 and R_5 support the normal condition diagnosis fully (100%), whereas the value of R_1 supports the condition with a certainty of 80% roughly. The percentage support (or belief) provided by each ratio is then averaged to give an overall support to that particular condition, in this case $(100 + 100 + 80)/3 = 93.3\%$ support to the Normal Condition case. A support for each fault scenario is calculated in this way, normalised using relative weightings and then combined using the original ER algorithm to produce an output of the following form[1]:

[1] This is the diagnoses set of DRM.

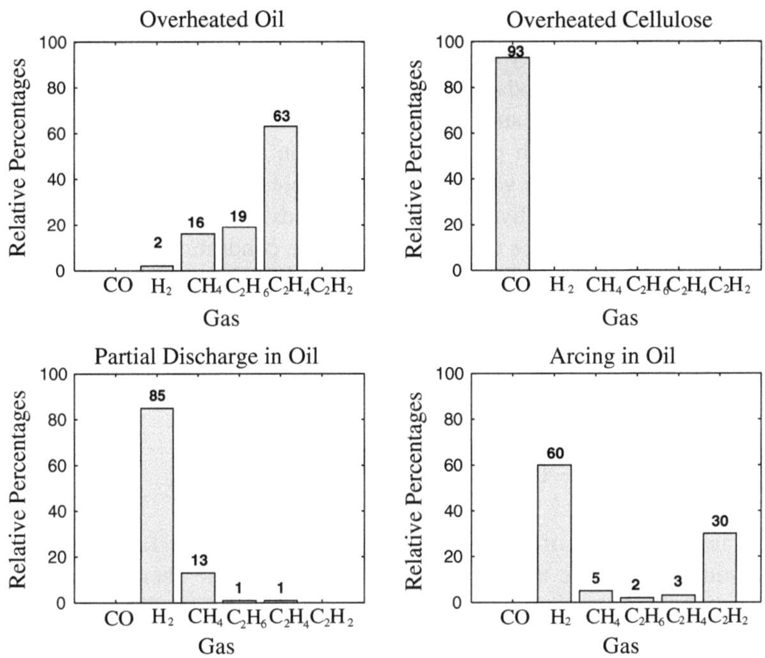

Fig. 8.5 Key gas method charts [4]

$$\text{Diagnosis} = \{\text{Pr(Thermal)}, \text{Pr(PD)}, \text{Pr(Arcing)}\}. \tag{8.14}$$

It must be noted that although there is no value for the possibility of the no fault condition, it is in fact implied since for all diagnoses:

$$\sum_{i=1}^{F_m} \text{Pr}(\text{Fault}_i) + \text{Pr(No Fault)} = 1, \tag{8.15}$$

where F_m is the number of fault scenarios determinable by a particular method, e.g. for DRM $F_m = 3$.

KGM does in fact not suffer from the same No Decision diagnosis problems as the two ratio methods. The reason for making the decision boundaries into fuzzy sets for this method is merely to provide results in the same format as those generated by the other methods.

8.3.2 Implementation of the Hybrid Diagnostic Approach

8.3.2.1 Problem Formulation with Evidential Reasoning

Often when dealing with decision making, a decision maker faces uncertainty. Such uncertainty can come from either a lack of knowledge about the problem, or uncertainty in the accuracy of the data used to make the decision. In the case of DGA, there is uncertainty in the accuracy of the diagnoses provided by the traditional diagnosis methods. What ER does is to provide a mathematical framework for combining such uncertain information (and subjective judgements). By considering each piece of information as evidence either supporting or denying a hypothesis, the validity of all possible hypotheses can be calculated. In the case of DGA, each hypothesis corresponds to a possible fault condition and the validity to the chance that this may be the condition of a transformer, e.g. 20% chance the transformer has suffered from or is currently suffering an arcing fault.

Before the ER algorithm can be used, there must be a formalism of the input data. ER requires that a set of common hypotheses are used for all sources of information H,

$$H = \{H_1, \ldots, H_n, \ldots, H_N\}, n = 1, \ldots, N, \tag{8.16}$$

where N is the number of hypotheses (in this case possible faults). So, a set of faults common to all the three methods must be devised. There are four general fault types distinguishable, OHC, OHO, PD and arcing, N.B. Thermal Faults are the same as OHO. Below are the fault types distinguishable by each method, Faults_R being the set of faults associated with RRM, Faults_D those associated with DRM and Faults_K with KGM:

$$\begin{aligned}\text{Faults}_\text{R} \in \{&\text{Low Temp}, \text{Thermal} < 700°\text{C},\\ &\text{Thermal} > 700°\text{C}, \text{PD}, \text{Arcing}\},\\ \text{Faults}_\text{D} \in \{&\text{Thermal}, \text{PD}, \text{Arcing}\},\\ \text{Faults}_\text{K} \in \{&\text{OHC}, \text{OHO}, \text{PD}, \text{Arcing}\}.\end{aligned}$$

It can be found from the three groups of fault types that by combining the supports from the subset {Low Temp, Thermal < 700°C, Thermal > 700°C} into the total support to a thermal fault and noticing that OHO is also in fact a thermal fault, the following minimum hypothesis set has been found to be exhaustive across all the three methods,[2]

$$H = \{\text{OHC, Thermal, PD, Arcing}\}. \tag{8.17}$$

Under an ER framework, a decision process is represented by a tree structure consisting of attributes, composite factors and basic factors. In this case the attribute is the overall evaluation of a transformer, i.e. its condition, which is supported by a set of composite factors comprising the results of the RRM diagnosis (RD), the DRM diagnosis (DD) and the KGM diagnosis (KGD). These composite factors are

[2] Pr(OHC) = 0 for both RRM and DRM as they cannot distinguish this fault directly.

in turn supported by three respective sets of basic factors, one for each DGA method. The role of the soft decision boundaries is to produce subjective judgements in the form of conditional probabilities, e.g. Pr(Fault|Gas), when presented with DGA data. An ER tree is developed based upon the above configuration as shown in Fig. 8.6.

Typically, the kth attribute of an ER tree is denoted as y_k. As in this case there is only one attribute as the overall evaluation, k is ignored throughout the forthcoming declaration. A set of composite factors $f = \{f_1, f_2, f_3\}$ represents the diagnoses made by each of the methods, i.e. $f = \{\text{RD, DD, KGD}\}$.

The fundamental difference between the tree proposed in this section and a typical ER tree (that is of the form presented in [7]) is that, the basic factors in this tree are not to be evaluated and they are in fact tools for generating subjective judgements. The set of basic factors can be grouped into three subsets $E = \{e_{RRM}, e_{DRM}, e_{KGM}\} = \{e_1, e_2, e_3\}$ where

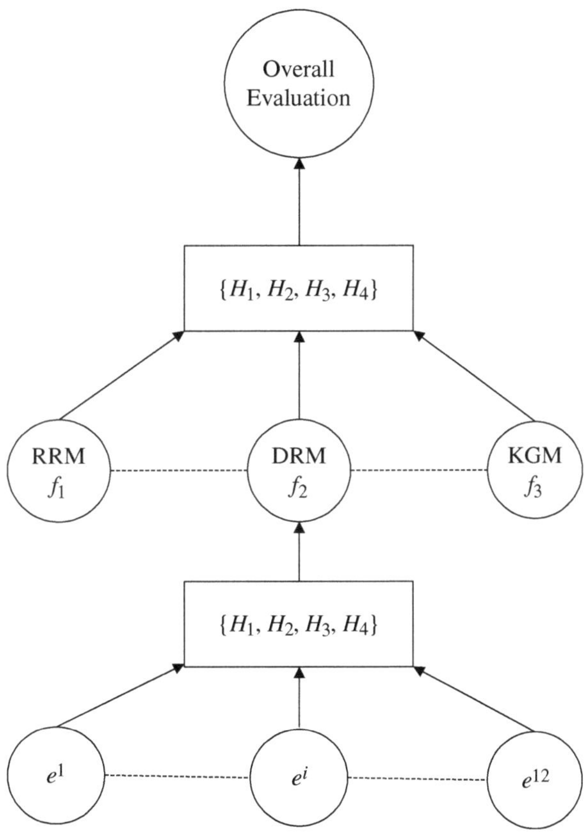

Fig. 8.6 An ER tree representing a DGA decision process

$$\begin{aligned} e_{\text{RRM}} &= \{\text{Low Temp., Thermal} < 700^\circ\text{C}, \\ &\quad \text{Thermal} > 700^\circ\text{C}, \text{PD}, \text{Arcing}\} \\ &= \{e_1^1, e_1^2, e_1^3, e_1^4, e_1^5\}, \\ e_{\text{DRM}} &= \{\text{Thermal}, \text{PD}, \text{Arcing}\} \\ &= \{e_2^1, e_2^2, e_2^3\}, \\ e_{\text{KGM}} &= \{\text{OHC}, \text{Thermal}, \text{PD}, \text{Arcing}\} \\ &= \{e_3^1, e_3^2, e_3^3, e_3^4\}. \end{aligned}$$

From this we can also define a set $L = \{L_1,L_2,L_3\}$, where L_k ($k = 1,2,3$) is a positive integer denoting the number of basic factors contributing to the evaluation of the composite factor f_k, so in this case $L = \{5,3,4\}$.

Unlike a typical ER basic factor set, these factors cannot be evaluated to all of the possible evaluation grades in H as each tool only generates the probability of one fault. Below are the groups of factors that contribute to the evaluation of each specific grade.

$$\begin{aligned} \{e_3^1\} &\to H_1, \\ \{e_1^1, e_1^2, e_1^3, e_2^1, e_3^2\} &\to H_2, \\ \{e_1^4, e_2^2, e_3^3\} &\to H_3, \\ \{e_1^5, e_2^3, e_3^4\} &\to H_4. \end{aligned}$$

Since each evaluation tool is deemed as accurate as the next, the relative weights for all of the basic factors are set to one, i.e. $\lambda = \{\lambda_1^1,...,\lambda_j^i,...,\lambda_3^4\} = \{1,1,...,1\}$. This in turn means that the set of normalised relative weights $\bar{\lambda}$ is also unity. To further simplify the problem, α, the coefficient representing the significance of the role of the most important factor (i.e. that with the highest weight), is also set to one. It should be noticed that the conditional probabilities generated by fuzzy sets represent the subjective judgements usually used as the inputs to an ER decision tree. What this means is that the conditional probability generated by factor e_k^i is denoted by $\beta_{H_n}(e_k^i)$, i.e. the belief with which factor e_k^i supports the evaluation of a transformer's condition to evaluation grade H_n. Before we can use the ER algorithm, such subjective judgements must be converted into basic probability assignments of the form $m_{k,i}^n$, where $m_{k,i}^n$ is the basic probability assignment generated by factor e_k^i that the transformer's condition is to be evaluated to grade H_n. To carry out this conversion, Eq. 8.18 must be used:

$$m_{j,i}^n = \bar{\lambda}_j^i \alpha \beta_{H_n}(e_j^i). \tag{8.18}$$

Since all $\bar{\lambda}$ and α equal 1, Eq. 8.18 is simplified as $m_{k,k}^n = \beta_{H_n}(e_k^i)$. With this in mind Tables 8.11, 8.12 and 8.13 are derived to illustrate the probability assignments of each composite factor.

A factor subset $e_{I_j(i)}$ is defined as

$$e_{I_k(i)} = \{e_k^1 \cdots e_k^i\}, 1 \leq i \leq L_k. \tag{8.19}$$

Table 8.11 Probability assignments for f_1

	H_1	H_2	H_3	H_4
e_1^1	0	$m_{1,1}^2$	0	0
e_1^2	0	$m_{1,2}^2$	0	0
e_1^3	0	$m_{1,3}^2$	0	0
e_1^4	0	0	$m_{1,4}^3$	0
e_1^5	0	0	0	$m_{1,5}^4$

Table 8.12 Probability assignments for f_2

	H_1	H_2	H_3	H_4
e_2^1	0	$m_{2,1}^2$	0	0
e_2^2	0	0	$m_{2,2}^3$	0
e_2^3	0	0	0	$m_{2,3}^4$

Table 8.13 Probability assignments for f_3

	H_1	H_2	H_3	H_4
e_3^1	$m_{3,1}^1$	0	0	0
e_3^2	0	$m_{3,2}^2$	0	0
e_3^3	0	0	$m_{3,3}^3$	0
e_3^4	0	0	0	$m_{3,4}^4$

Table 8.14 Intermediate evaluation table

	H_1	H_2	H_3	H_4
f_1	m_1^1	m_1^2	m_1^3	m_1^4
f_2	m_2^1	m_2^2	m_2^3	m_2^4
f_3	m_3^1	m_3^2	m_3^3	m_3^4

We can also define a combined probability assignment $m_{I_k(i)}^{\Psi}$, i.e. the probability that the transformer's condition is to be evaluated to $\Psi(\Psi \subseteq H)$, supported by the factor subset $e_{I_k(i)}$. The last term to be defined is $m_{k,i}^H$ as the remaining belief not assigned after commitment to all $H_n(n = 1,...,N)$, i.e. $m_{k,i}^H = 1 - \sum_{n=1}^N m_{k,i}^n$.

With this formalism in place and by stating that $e_{I_k(1)} = e_k^1$, the original ER algorithm in Eqs. 8.7–8.9 is used to combine the evidence presented by the simulated subjective judgements at the basic factor level.

Let $m_{I_k(L_k)}^n = m_k^n$, then an intermediate evaluation Table 8.14 is generated.

By defining a composite factor subset $f_{I(i)} = \{f^1,...,f^i\}$ in the same vein as $e_{I_k(i)}$, then the probabilities in Table 8.14 can be combined using a slightly simpler form of the ER algorithm stated in Eqs. 8.20–8.22.

For $i = 1,2$

$$\{H_n\} : m_{I(i+1)}^n = K_{I(i+1)}(m_{I(i)}^n m_{i+1}^n + m_{I(i)}^n m_{i+1}^H + m_{I(i)}^H m_{i+1}^n), \quad n = 1,\ldots,N, \tag{8.20}$$

$$\{H\} : m_{I(i+1)}^H = K_{I(i+1)} m_{I(i)}^H m_{i+1}^H, \tag{8.21}$$

$$K_{I(i+1)} = \left[1 - \sum_{\tau=1}^{N} \sum_{j=1, j \neq \tau}^{N} m_{I(i)}^{\tau} m_{i+1}^{j}\right]^{-1}. \tag{8.22}$$

This final set of equations now presents us with an overall evaluation of transformer conditions, in the form shown in Eq. 8.17, based on the evidence provided by the three traditional DGA methods altered to produce pseudo subjective judgements. The proof and theory supporting this algorithm are reported extensively in [7] and [8].

8.3.2.2 Fuzzy Membership Functions of the Gas Ratio Methods

The first step in implementing fuzzy sets is to find suitable membership functions to soften the decision boundaries used by the gas ratio methods. Equation 8.23 shows a sigmoidal function employed in this study to soften crisp decision boundaries.

$$f(x) = (1 + e^{-a(x+c)})^{-1}. \tag{8.23}$$

The slope a of the sigmoid function and the value on which the functions are cantered c have been derived by inspecting the Halstead's diagram (Fig. 8.4), by analysing the relationships between the gases involved in each ratio with respect to temperatures.

Where rules state that the value of a ratio must lie within a specific range, a Gaussian bell function of the form in Eq. 8.24 is used to soften the boundary. Again the function parameters a, b and c are derived from the Halstead's diagram, c having the same meaning as before whilst a determines the width of the bell and b the steepness of the bell's edges.

$$f(x) = e^{((x-c)/a)^{2b}}. \tag{8.24}$$

With the soft boundaries now in place, the support for each fault type, provided by the individual gas ratios, is required to be normalised. For simplicity it is assumed that each ratio is equally important for every fault type diagnosis. What this means practically is that the overall support for each fault type is simply the average of the supports provided by each ratio. The overall supports for each fault are then passed into the ER algorithm described in Sect. 3.2.1 and an evaluation of all possible transformer faults, as distinguishable by that method, is produced. For example, considering the DGA data in Table 8.15, these data are processed

Table 8.15 Example DGA data

Gas	H_2	CH_4	CO	C_2H_2	C_2H_4	C_2H_6
ppm	270	190	280	37	17	4

producing the following diagnoses concerning the new DRM. The value of R_1 provides 92.7% support for there being a thermal fault, i.e.

$$\begin{aligned}\Pr(\text{Thermal}|R_1)_{\text{DRM}} &= 0.9270,\\ \Pr(\text{Thermal}|R_2)_{\text{DRM}} &= 0.0790,\\ \Pr(\text{Thermal}|R_3)_{\text{DRM}} &= 0.9988,\\ \Pr(\text{Thermal}|R_4)_{\text{DRM}} &= 0.5000.\end{aligned}$$

These values are then averaged as described above to give an overall support of 62.62% to the thermal fault condition. Similarly:

1. Support for a PD fault = 41.31%.
2. Support for an arcing fault = 73.13%.

The final step is to combine these three pieces of evidence using the ER algorithm detailed in Sect. 3.2.1 to provide us with an overall diagnosis by the new DRM. It must be noted that the support values for each fault type also imply a support for all other possible diagnoses including the no fault condition, and this is why the results from the ER combination are not merely a normalisation of the three basic supports calculated previously. As a result, the overall evaluation using ER[3,4] is [0, 0.235, 0.434, 0.114].

From these results we can also work out that the remaining belief is $1 - 0.235 - 0.434 - 0.114 = 0.217$, i.e. a 21.7% chance of other possible faults or no fault having occurred. Using the same principles and practices, a similar version of the new RRM can also be implemented. With both the ratio methods with fuzzy sets now in place, the next step is to try and apply the same ideas to KGM.

8.3.2.3 Fuzzy Membership Functions of the Key Gas Method

The difference between KGM and the ratio methods is that KGM requires an engineer to make a subjective decision on the level of correlation between the DGA data and the four fingerprint charts presented in Sect. 8.1. Such a judgement would be of the kind: "I believe that the data I have matche the OHC characteristic chart, and I can say this with 50% certainty (assurance)." Obviously such decision making is not easily put into mathematics, however if we break down the judgement into a set of thought processes then it becomes much easier. Consider the DGA data presented in Table 8.15, and the relative percentages are shown in Table 8.16.

These values should be compared with those presented in each of the four diagrams shown in Fig. 8.5. Only the arcing and OHC diagrams are discussed here to illustrate the ideas behind the derivation of the fuzzy sets.

---

[3] Refer back to Eq. 8.14 for the set of fault types.

[4] The zero represents the probability of OHC, since DRM cannot determine this fault.

Table 8.16 Example DGA data (relative percentages)

Gas	H_2	CH_4	CO	C_2H_2	C_2H_4	C_2H_6
%	34	23	35	5	2	1

The key gas generated by an arcing fault is in fact C_2H_2, not H_2 as the chart may first suggest. This is because C_2H_2 is only produced at temperatures in excess of 1,000°C, which is a temperature only found when arcing occurs. So, an engineer's first thought should be: "Is there a significant percentage of C_2H_2?". To answer this the engineer must first decide what is meant by a significant percentage. This choice is one that typically should be made by an experienced DGA reviewer based on their own experiences, e.g. 20% is deemed the absolute minimum value of C_2H_2 that would indicate an arcing fault. Hence the probability of arcing given the percentage of C_2H_2 is found using a sigmoidal function similar to those used to soften the decision boundaries of the ratio methods, with $a = 0.15$ and $c = 27$.

Similar curves are used for all of the other gases to generate conditional fault probabilities, e.g. $\Pr(\text{Arcing}|H_2\%) \equiv$ Probability of Arcing given the relative percentage of H_2. These conditional probabilities are to be combined to generate the overall probability of an arcing fault given the relative amounts of all the gases. When combining these probabilities, it should be noted that the key features in the arcing fault chart are:

1. Significant presence of C_2H_2.
2. Large quantity of H_2 with respect to CO, CH_4,C_2H_6 and C_2H_4.

To ensure that these features are dominant over the other gas concentrations, the normalised probabilities of all the gases except H_2 and C_2H_2, are multiplied by a scaling factor K_{OHC}. The probabilities are first normalised by simply dividing each one by 6, i.e. the number of gases, therefore the overall arcing probability lies in the closed interval [0, 1]. The scaling factor used is actually the average of the conditional probabilities of arcing given H_2 and C_2H_2 levels.

$$K_{\text{OHC}} = \frac{\Pr(\text{Arcing}|\text{H}_2) + \Pr(\text{Arcing}|\text{C}_2\text{H}_2)}{2}. \tag{8.25}$$

The choice of this scaling factor ensures that the overall probability of an arcing fault is only close to 1, when both the levels of H_2 and C_2H_2 are close to those prescribed in the key gas chart (Fig. 8.5). Using both the scaling factor and normalised probabilities, the overall probability of an arcing fault given the relative percentages of all the fault gases is shown in Eq. 8.26. It should be noticed that the weights allocated to each probability are selected for an illustration purpose.

$$\Pr(\text{Arc}) = \left(\frac{\Pr(\text{Arc}|\widetilde{\text{A}})}{6}\frac{\Pr(\text{Arc}|\widetilde{\text{B}})}{2}\right) + \frac{\Pr(\text{Arc}|\widetilde{\text{B}})}{6}, \tag{8.26}$$

where

$$\begin{aligned}
\Pr(\text{Arc}) &= \Pr(\text{Arcing})_{\text{KGM}},\\
\widetilde{A} &= \{\text{CO}, \text{CH}_4, \text{C}_2\text{H}_6, \text{C}_2\text{H}_4\},\\
\widetilde{B} &= \{\text{H}_2, \text{C}_2\text{H}_2\},\\
\Pr(\text{Fault}|\text{Gas}_1, \ldots, \text{Gas}_N) &= \sum_{i=1}^{N} \Pr(\text{Fault}|\text{Gas}_i).
\end{aligned}$$

Similarly, when comparing the relative percentages of DGA gases against the OHC chart shown in Fig. 8.5, the key feature is identified as the large presence of CO. This is because the only source of O_2 within the air-tight, oil-insulated part of the transformer is the paper insulation often found around windings and in contact with the oil. Since the presence of CO in any great abundance is the sign of cellulose overheating, a similar technique as above is used to scale the conditional probabilities generated by the levels of the other gases. Whereas the average of the conditional probabilities generated by H_2 and C_2H_2 are used for arcing faults, simply the normalised conditional probability generated by CO is used for this case. This means that the overall probability of an OHC fault given the relative percentages of all the fault gases is calculated using Eq. 8.27, where the shorthand notation Pr(Fault|Gas_1, Gas_2,..., Gas_N) has the same meaning as before.

$$\Pr(\text{OH}) = \left(\frac{\Pr(\text{OH}|\text{CO})}{4}\frac{\Pr(\text{OH}|\widetilde{\text{A}})}{5}\right) + \frac{3\Pr(\text{OH}|\text{CO})}{4}, \tag{8.27}$$

where

$$\begin{aligned}
\Pr(\text{OH}|\text{Gas}) &= \Pr(\text{OHC}|\text{Gas})_{\text{KGM}},\\
\Pr(\text{OH}) &= \Pr(\text{OHC})_{\text{KGM}},\\
\widetilde{A} &= \{\text{H}_2, \text{CH}_4, \text{C}_2\text{H}_6, \text{C}_2\text{H}_4, \text{C}_2\text{H}_2\}.
\end{aligned}$$

The choice of scaling factors ensures that the overall probability of an OHC fault is only close to 1, when the level of CO is close to that prescribed in the key gas chart (Fig. 8.5). Notice that because the relative value of CO is paramount to there being an OHC fault, the relative weights are set as [0.75, 0.05, 0.05, 0.05, 0.05, 0.05]. If considering the example data shown in Table 8.15, the following conditional probabilities are generated for an arcing fault:

$$\begin{aligned}
\Pr(\text{Arcing}|\text{CO}) &= 0.0000,\\
\Pr(\text{Arcing}|\text{H}_2) &= 0.1634,\\
\Pr(\text{Arcing}|\text{CH}_4) &= 0.0000,
\end{aligned}$$

$$\Pr(\text{Arcing}|C_2H_6) = 0.9999,$$
$$\Pr(\text{Arcing}|C_2H_4) = 0.9996,$$
$$\Pr(\text{Arcing}|C_2H_2) = 0.0337.$$

Using Eq. 8.26 it is found the probability of an arcing fault to be $\Pr(\text{Arcing})_{KGM} = 0.0657$, i.e. 6.57%. Similarly, if the OHC method is employed using the new KGM technique, the following conditional probabilities are generated:

$$\Pr(\text{OHC}|\text{CO}) = 0.5044,$$
$$\Pr(\text{OHC}|H_2) = 0.0000,$$
$$\Pr(\text{OHC}|CH_4) = 0.0121,$$
$$\Pr(\text{OHC}|C_2H_6) = 0.9993,$$
$$\Pr(\text{OHC}|C_2H_4) = 0.9984,$$
$$\Pr(\text{OHC}|C_2H_2) = 0.9944.$$

Using Eq. 8.27 it is found the probability of an OHC fault to be $\Pr(\text{OHC})_{KGM} = 0.4541$. This process is carried out again for all the other types of faults that KGM can distinguish between, namely Thermal faults (overheating of oil) and PD faults. The overall probabilities of these faults are shown as below:

$$\Pr(\text{Thermal})_{KGM} = 0.0386,$$
$$\Pr(\text{PD})_{KGM} = 0.3754.$$

With all the four faults now having an associated probability, the values can be put into the original ER algorithm and the following results generated. The overall evaluation using the original ER algorithm[5] is [0.327, 0.0158, 0.2364, 0.0277]. These results also imply a probability of no fault, i.e. $\Pr(\text{No Fault})_{KGM} = 0.3931$.

8.3.3 *Tests and Results*

Using the data in Table 8.15, the evaluations shown in Tables 8.17, 8.18 and 8.19 are made by the three traditional DGA methods.

Combining these probabilities using the original ER algorithm gives the intermediate evaluation as shown in Table 8.20.

The ER algorithm in Eqs. 8.20–8.22 is now used to combine the evidence in the intermediate table and gives an overall evaluation of the transformer's condition as listed in Table 8.21.

---

[5] Refer back to Eq. 8.14 for the set of fault types.

Table 8.17 Conditional probabilities for the Rogers ratio method

	H_1	H_2	H_3	H_4
e_1^1	0	0.3287	0	0
e_1^2	0	0.1072	0	0
e_1^3	0	0.1072	0	0
e_1^4	0	0	0.3438	0
e_1^5	0	0	0	0.3228

Table 8.18 Conditional probabilities for the Dörnenburg's ratio method

	H_1	H_2	H_3	H_4
e_2^1	0	0.6262	0	0
e_2^2	0	0	0.4131	0
e_2^3	0	0	0	0.7313

Table 8.19 Conditional probabilities for the key gas method

	H_1	H_2	H_3	H_4
e_3^1	0.4541	0	0	0
e_3^2	0	0.0386	0	0
e_3^3	0	0	0.3754	0
e_3^4	0	0	0	0.0657

Table 8.20 Intermediate evaluation table

	H_1	H_2	H_3	H_4
f_1	0	0.2673	0.1919	0.1746
f_2	0	0.2746	0.1154	0.4461
f_3	0.3270	0.0158	0.2364	0.0277

Table 8.21 Final overall evaluation table

H_1	H_2	H_3	H_4	No fault
0.0608	0.2787	0.2308	0.3567	0.0731

What this illustrates is that there is a 35.67% chance that the transformer is suffering from or has suffered from an arcing fault, a 27.87% chance that the fault is a Thermal fault, a 23.08% chance that the fault is a PD fault and a 6.08% chance that the fault is cellulose degradation. Given these results and the fact that the chance of no fault having occurred is only very small (7.31%), then it is more than likely that the transformer in question has suffered from an arcing fault. This diagnosis does in fact match the official diagnosis taken by NG engineers made after the failure of the transformer in question, i.e. "Failed due to arcing between the insulated OLTC shaft pin and the coupling of drive".[6]

[6] Diagnosis taken from the NG DGA database.

When these results are compared with those produced by the traditional DGA methods, the advantages of the new system become apparent. The Rogers ratio method finds an arcing fault, which although correct fails to convey the extra information concerning the probable overheating of cellulose that has also been occurring or has previously occurred. The Dörnenburg's method produces the No Decision diagnosis and the inspection of the key gas method is inconclusive without extra information. The key factor in the representation of the results from the new system is the fact that the probability of the no fault case is very small, clearly highlighting a danger to the transformer, whilst the presence of a previous cellulose overheating problem and the correct diagnosis of the arcing fault are also presented, therefore helping an engineer understand more clearly what has occurred inside the transformer. This comparison highlights the nature of the improvement offered by this hybrid diagnostic approach over the three traditional DGA methods.

8.4 Probabilistic Inference Using Bayesian Networks

As stated previously, when DGA reviewers examine gas ratio data, they evaluate the gas ratios that they have according to decision rules of several traditional DGA guidelines. Subjective decisions are then made, and a DGA reviewer can check whether the data in hand fit any of the decision criteria or not. The diagram of a typical DGA fault diagnosis process is shown in Fig. 8.7, which includes different gas ratios, fault types and fault diagnosis actions.

It is known that, not all the combinations of gas ratios presented in a fault can be mapped to a fault type in a selected criterion, e.g. the Rogers ratio method. In this case, a DGA reviewer has to decide how close are the data to the chosen criterion. Considerations have been given to the interpretation of such combinations of gas ratios, as some combinations of gas ratios presented in certain types of faults are not reported in the IEEE and IEC codes for DGA interpretation [6]. In other words, it is a puzzling task of how to handle a fault scenario where gas ratio combinations are missing in the IEEE and IEC DGA codes. Many attempts have since been made to refine a decision process used to guide DGA reviewers, such attempts include data analysis using EPS [9] and ANN [10]. However, such attempts are limited in their representation of the problem as a pattern recognition task, which are defined as black-box models with only inputs and outputs. The logical reasoning process of a DGA reviewer is not mimicked, which is essentially a probabilistic inference process. The probabilistic reasoning approach detailed in this section is to enhance DGA fault diagnosis capabilities along with traditional fault classification criteria by constituting a BN for DGA diagnosis problems.

For a DGA reviewer, the core tasks of transformer fault diagnosis are to identify the relationships between fault gas ratios and fault types. It is essentially a probabilistic reasoning process to compute the unknown (posterior) probabilities of a

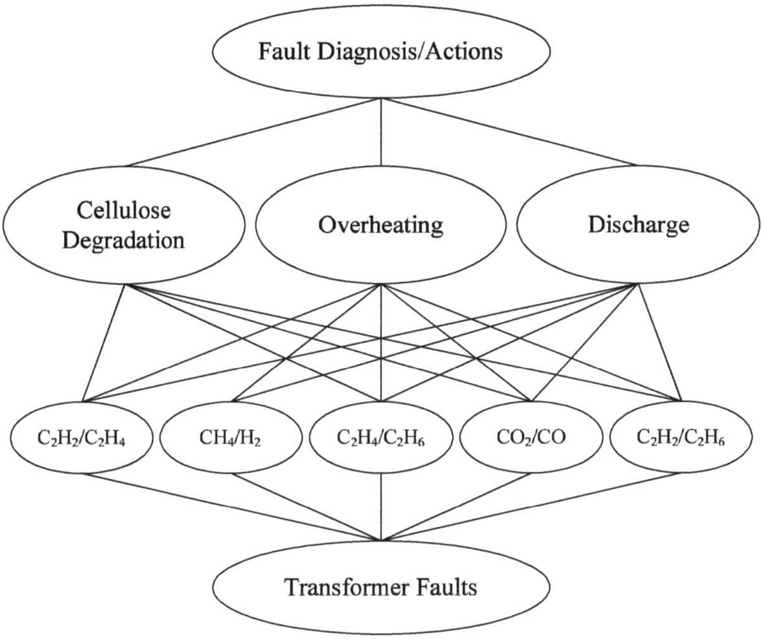

Fig. 8.7 A typical decision diagram of DGA fault diagnosis

certain type of fault, given new evidence as groups of key gas ratios based upon known (prior) probabilities. In this sense, if a DGA problem is regarded as a decision making process under the framework of probability theories, transformer fault diagnosis can be treated as a probability inference task. If a BN is employed as a framework for conducting DGA, by creating a BN according to the diagram of DGA interpretation in Fig. 8.7, a probability inference procedure can be implemented to identify fault types of a transformer based on the evidence of various combinations of gas ratios.

8.4.1 Knowledge Transformation into a Bayesian Network

In order to create a BN for performing transformer fault diagnosis, the major modelling issues that arise are:

1. What are the variables/nodes?
2. What is the graph structure?
3. What are the parameters (CPT)?

In this subsection, the procedures of knowledge transformation for eliciting a BN structure and its parameters are illustrated step by step to obtain accurate DGA interpretations.

8.4.1.1 Determination of BN Variables Based on the IEEE and IEC Codes for DGA Interpretation

When attempting to model transformer fault diagnosis problems, it is important to determine the number of varialbes/nodes before establishing a BN with a concise and appropriate structure. As known, the main interest of transformer fault diagnosis is to identify fault types of a unit based upon gas ratios. Typical faults are thermal faults, discharge faults and faults involving cellulose degradation as shown in Fig. 8.7. An initial BN modelling choice is to have two query nodes of fault types, with the observation nodes being three sets of gas ratios from the IEEE and IEC DGA coding schemes [3], which are depicted in Fig. 8.8.

When selecting BN variables, we must also decide what states or values the BN variables can take. As the purpose of this study is to develop an approach to enhancing the IEEE/IEC fault diagnosis criteria, the values of BN variables are extracted from the IEEE/IEC codes for interpretation of DGA results, which are briefly introduced below.

As shown in Table 6.1, the gas ratio methods are coding systems that assign a certain combination of codes to specific fault types. The codes are generated by calculating ratios of gas concentrations and comparing the ratios to predefined limits, which were derived from experiments and industrial experience. A fault condition is detected when a code combination fits the code pattern of a fault. The most commonly used ratio method is the Rogers ratio method [6], which is able to distinguish more types of thermal faults than the Dörnenberg's ratio method, listed in Table 6.1 [3]. In this study, the values of nodes exhibiting fault types are chosen as [Normal, Low temperature overheating (LT-H), High temperature overheating (HT-H)] and [Normal, Low energy discharge (LE-D), High energy discharge (HE-D)] for the nodes of "overheating fault" and "discharge fault" respectively. The types of node values for "gas ratios" are selected as the discrete code patterns defined in Table 6.1, e.g. [0, 0, 0] illustrate a normal condition.

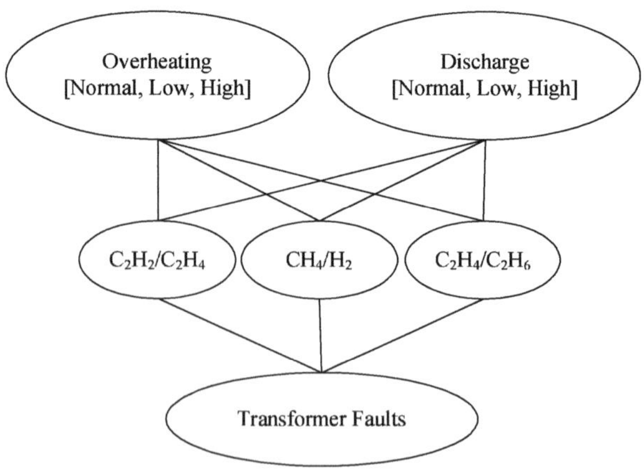

Fig. 8.8 A DGA reasoning network based on IEEE/IEC DGA interpretation codes

While applying the gas ratio method in accordance with Table 6.1, it should be noted that:

1. The ratios for a combination of multiple faults may not fit the predefined codes in Table 6.1.
2. Combinations of ratios, not included in Table 6.1, may occur in practice. Consideration is being given to the interpretation of such combinations.

The proposed BN approach is developed to overcome the drawbacks of missing codes in Table 6.1 for accurate transformer fault classifications.

8.4.1.2 Construct a Graphical Model with IEEE and IEC Codes

When deciding on the structure of a BN, the key is to focus on relationships between variables. By studying the reasoning process of the IEEE and IEC DGA coding schemes, it is clear that a DGA fault diagnosis problem is actually a belief reasoning process based upon a set of evidence. The DGA reasoning diagram in Fig. 8.8 can then be easily transformed into a graphical model shown in Fig. 8.9, i.e. a BN.

As shown in Fig. 8.9, the root nodes of X_1 and X_2 represent two fault types listed in Table 6.1, i.e. overheating fault and discharge fault. In this study, the fault types in Table 6.1 are reorganised into two groups, i.e. X_1 possesses three states [Normal, LT-H, HT-H] and X_2 also represents three states [Normal, LE-D, HE-D]. The leaf nodes of X_3, X_4 and X_5 represent three types of ratios, which are $\frac{C_2H_2}{C_2H_4}$, $\frac{CH_4}{H_2}$ and $\frac{C_2H_4}{C_2H_6}$, respectively. The possible states of each leaf node are defined in Table 6.1, e.g. [0, 1, 0] and [1, 1, 0], which are determined by fault gas quantities and types.

8.4.1.3 Further Explanation on Bayesian Network Inference

In Fig. 8.9, a directed arc graph (DAG) is depicted, where all arcs (or arrows) point downwards. The directions of these arcs represent relationships of causes and

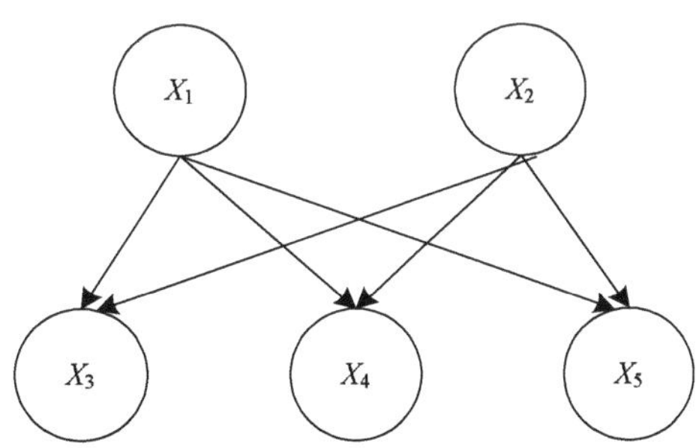

Fig. 8.9 A simple Bayesian network for DGA analysis

effects occurring in DGA based fault diagnosis, e.g. a high energy discharge fault results in a high ratio of $\frac{C_2H_2}{C_2H_4}$.

For each node in Fig. 8.9, a CPT quantifies the effects parents have on a child node. CPT values can be obtained from historical fault pattern data by automated machine learning. As the CPT of each node is unknown when creating a BN, random parameters are generated from a uniform distribution. By now, a BN for analysing dissolved gases is created, in which the data structure represents the dependance between the variables of all the nodes. When the construction of a BN is completed, it is ready to deploy the developed model to tackle real-world DGA fault diagnosis problems.

8.4.1.4 Implementation the Bayesian Network Approach to DGA Interpretation

The following procedures describe how to build a BN from a practitioner's viewpoint, which are implemented in the next subsection.

1. Design a graphical model for representing a DGA fault classification problem, including both variables and the data structure of a specified BN.
2. Collect a group of DGA data sampled from faulty transformers, which include both key gas quantities and actual fault types with the same form as these defined in Table 6.1.
3. Determine the CPT for each node in the established BN, by using an automated parameter learning programme for BN construction.
4. Probabilistic inference for fault classifications: compute the probability distribution for query variables (X_1 and X_2) given evidence variables (X_3, X_4 and X_5).
5. Update the CPTs of the constructed BN classifier with new fault cases if necessary.

8.4.2 Results and Discussions

In this section, the BN approach is utilised to undertake transformer fault diagnosis tasks based upon DGA data of a group of transformers. The "Bayes net toolbox for MATLAB" written by Kevin Murphy is employed for this study.[7] The test DGA data and matching diagnoses are taken directly from the NG DGA database. The procedures of BN construction and inference for a transformer DGA problem are illustrated as follows:

[7] http://www.code.google.com/p/bnt/

1. Firstly, 50 sets of DGA data are extracted from the NG DGA database, which contain both the 7 types of key gases and the diagnosis results after on-site inspections conducted on a group of transformers removed for repair. The datasets are then divided into two groups, i.e. datasets 1 and 2.
2. Dataset 1 comprises 40 sets of records, which are used for CPT parameter learning with regard to the created BN depicted in Fig. 8.9. Through a BN learning programme, the CPT of each node is ascertained based upon dataset 1.
3. The remaining 10 sets of records form dataset 2, which is employed for the evaluation of the created BN with the derived CPT values. It is noted that the fault types of five cases in dataset 2 cannot be identified using the IEEE and IEC DGA coding scheme due to missing codes.

8.4.2.1 Derivation of CPTs from Training Data

For illustration purposes, in this subsection the derivation of CPTs of X_3, X_4 and X_5 are presented for discharge fault analysis. By feeding DGA dataset 1 through a BN parameter learning programme, the derived CPTs of nodes X_3, X_4 and X_5 regarding node X_2 are listed in Tables 8.22, 8.23 and 8.24 respectively. Each row in the tables contains the conditional probability of each node for each possible combination of values of its parent node X_2, i.e. relationships between two sets of predefined categories—possible gas ratio codes of 0, 1 or 2 and transformer states as [Normal, LE-D, HE-D]. When the graph structure and CPTs of all the nodes of the BN are identified, it is ready to deploy the BN to perform transformer fault diagnosis tasks.

Table 8.22 Conditional probability table for X_3 regrading X_2

	Normal	LE-D	HE-D
0	0.57	0.43	0
1	0.40	0.50	0.10
2	0	0.95	0.05

Table 8.23 Conditional probability table for X_4 regrading X_2

	Normal	LE-D	HE-D
0	0.65	0.09	0.26
1	0.15	0.75	0.10
2	0.85	0.05	0.10

Table 8.24 Conditional probability table for X_5 regrading X_2

	Normal	LE-D	HE-D
0	0.48	0.43	0.09
1	0.75	0.05	0.20
2	0	0.07	0.83

8.4.2.2 Validation of the Constructed Bayesian Network

In order to verify the effectiveness of the constructed BN, dataset 2 is employed to carry out probabilistic inference to identify possible fault types, which are then compared with actual diagnosis from on-site inspections.

For the first five sets of cases 1 to 5 in Table 8.25, the IEEE and IEC DGA codes and on-site diagnosis results are listed in Table 8.26.

By feeding the data of the five cases to the constructed BN model, the diagnosis results using probabilistic inference are shown in Table 8.27.

As illustrated in Table 8.27, for cases 1 to 5 in Table 8.25, the diagnosis results obtained from the IEEE and IEC DGA coding scheme are the same as the diagnosis results derived from the BN inference. For instance, as defined in Sect. 8.4.1.2, the states of nodes X_1 and X_2 are the probabilities of [Normal, LT-H, HT-

Table 8.25 DGA concentrations of 10 scrapped transformers (in ppm)

Gas	H_2	CH_4	C_2H_2	C_2H_4	C_2H_6
Case 1	33	26	0.2	5.3	6
Case 2	21	124	5	183	45
Case 3	835	76	16	10	29
Case 4	1,570	1,110	1,830	1,780	175
Case 5	3,420	7,870	33	6,990	1,500
Case 6	833	3,167	1,697	5,793	390
Case 7	217	286	884	458	14
Case 8	41	112	4,536	254	0
Case 9	95	10	39	11	0
Case 10	60	5	21	21	2

Table 8.26 IEEE and IEC codes and on-site inspections for cases 1–5

	IEEE/IEC codes	On-site inspections
Case 1	[0, 0, 0]	Normal
Case 2	[0, 2, 2]	HT-H
Case 3	[1, 1, 0]	LE-D
Case 4	[1, 0, 2]	HE-D
Case 5	[0, 2, 2]	HT-H

Table 8.27 Comparisons between IEEE/IEC and Bayesian network diagnosis for cases 1 to 5

	IEEE/IEC	Prob. of X_1, X_2	BNs
case 1	Normal	X_2 [**0.82**, 0.18, 0.00]	Normal
case 2	HT-H	X_1 [0.02, 0.04, **0.94**]	HT-H
case 3	LE-D	X_2 [0.07, **0.93**, 0.00]	LE-D
case 4	HE-D	X_2 [0.02, 0.01, **0.97**]	HE-D
case 5	HT-H	X_1 [0.02, 0.04, **0.94**]	HT-H

BNs Bayesian networks, *Prob.* probabilities belong to a node (X_1 or X_2) of the created BN. The bold values indicate the maximum values among the three elements of the probability vectors of x_1 and x_2.

H] and [Normal, LE-D, HE-D] respectively. With regard to case 1, the IEEE/IEC interpretation provides a code of [0, 0, 0], which is classified as a normal condition. By inputting the data of case 1 to the created BN inference engine, the derived probability of node X_2 is [0.82, 0.18, 0.0], which also indicates a high probability of no fault. As for cases 2 and 5, there are high probabilities of HT-H, i.e. 0.94 and 0.94, which are the same as the IEEE/IEC DGA interpretations. Moreover, it is indicated in cases 3 and 4 that high probabilities of LE-D and HE-D, i.e. 0.93 and 0.97, also coincide with the IEEE/IEC DGA interpretations. The developed results show that the constructed BN model can successfully map a group of DGA codes to certain types of faults, which are also identifiable by the IEEE/IEC DGA coding scheme.

For cases 6–10 in Table 8.25, the IEEE/IEC codes and on-site diagnosis results are listed in Table 8.28. It should be noted that the IEEE/IEC DGA interpretations cannot be implemented, as the code combinations of the 5 cases are not mentioned in Table 6.1.

It is expected that, by feeding the 5 cases to the constructed BN, the fault classification can be implemented while the codes of cases 6–10 are missing in the IEEE/IEC DGA coding scheme. The diagnosis results from the BN inference model are shown in Table 8.29, which coincide with the actual on-site diagnosis. For example, in case 6, the IEEE/IEC coding scheme provides a code of [1, 2, 2], which cannot been classified due to missing codes in Table 6.1. By inputting the DGA data of case 6 to the developed BN, the derived probability of node X_1 is [0.08, 0.01, 0.91], which indicates a high probability of a fault with HT-T. For case 10, the probabilities delivered by BN diagnosis are 0.47 and 0.50 for fault types of LE-D and HE-D respectively, bearing in mind that the code of [1, 1, 2] is missing in the IEEE/IEC DGA coding scheme. The results cannot be

Table 8.28 IEEE/IEC codes and on-site inspections for cases 6 to 10

	IEEE/IEC codes	On-site inspections
Case 6	[1, 2, 2]	HT-H
Case 7	[1, 2, 2]	HT-H
Case 8	[2, 2, 2]	HE-D
Case 9	[2, 1, 2]	HE-D
Case 10	[1, 1, 2]	LE-D

Table 8.29 Comparisons between IEEE/IEC and BN diagnosis results for cases 6–10

	IEEE/IEC	Prob. of X_1, X_2	BNs
Case 6	N/A	X_1 [0.08, 0.01, **0.91**]	HT-H
Case 7	N/A	X_1 [0.08, 0.01, **0.91**]	HT-H
Case 8	N/A	X_2 [0.00, 0.20, **0.80**]	HE-D
Case 9	N/A	X_2 [0.01, 0.11, **0.89**]	HE-D
Case 10	N/A	X_2 [0.03, 0.47, **0.50**]	HE-D

N/A The diagnosis code of each case is not available in the IEEE/IEC DGA coding scheme. The bold values indicate the maximum values among the three elements of the probability vectors of x_1 and x_2.

Table 8.30 Comparisons between IEEE/IEC and BN diagnosis

Codes	IEEE/IEC	Prob. of X_2	BNs
[0, 0, 0]	Normal	X_2 [**0.82**, 0.18, 0.00]	Normal
[0, 1, 0]	LE-D	X_2 [0.11, **0.89**, 0.00]	LE-D
[1, 1, 0]	LE-D	X_2 [0.07, **0.93**, 0.00]	LE-D
[1, 0, 1]	HE-D	X_2 [0.33, 0.01, **0.66**]	HE-D
[1, 0, 2]	HE-D	X_2 [0.02, 0.01, **0.96**]	HE-D
[2, 0, 2]	HE-D	X_2 [0.00, 0.04, **0.96**]	HE-D

The bold values indicate the maximum values among the three elements of the probability vectors of x_1 and x_2.

further classified into the faults types of LE-D or HE-D due to very close probabilities, which can still indicate a trend of discharge fault.

8.4.2.3 Verification of the IEEE and IEC DGA Coding Scheme

In Table 8.30, the marginal probabilities of X_2 are calculated providing the codes extracted from Table 6.1 concerning discharge faults.

It can be noted from Table 8.30 that the diagnosis from the IEEE/IEC DGA codes can always be mapped correctly to the maximum values of the marginal probabilities of node X_2, regarding the same code combinations extracted from Table 6.1. The comparisons between the BN outputs and the IEEE/IEC diagnosis are consistent to each other, which can prove the effectiveness of the IEEE/IEC DGA coding scheme to some extent and further validate the proposed BN approach.

8.4.2.4 Discussion

The main methodology and advantages of the proposed BN approach are summarised as below:

1. The IEEE/IEC DGA coding scheme can be mapped directly into a BN solution.
2. The derived CPTs of each node are the conditional probabilities to its parent nodes. Compared with weights and biases for ANN modelling regarding DGA problems, the CPT representation of each node is more meaningful and intuitive. The derived probabilities can greatly help a DGA reviewer identify conditions of a transformer in question with a clear meaning under a firm mathematical foundation. While using an ANN method, only crisp diagnosis (either fault or no fault) is produced and the weights of a derived neural network have no physical meanings.
3. The validity of the IEEE/IEC DGA interpretation codes has been proven by calculating the marginal and joint probability distributions of each type of faults in Table 8.30. Moreover, the cases, which are unidentifiable by the IEEE/IEC

DGA interpretation scheme due to missing codes, can be successfully identified as shown in Table 8.29.
4. Finally, the developed BN approach can be expanded easily to a more complex graph structure, which is applicable for a variety of engineering applications not limited in this particular DGA problem.

8.5 Summary

In the first part of this chapter, an ER approach has been developed to assess the condition of a power transformer, as well as to rank conditions of a group of transformers for power system maintenance purposes. The methodology of transferring a transformer condition assessment problem into an MADM solution under an ER framework is presented. Several solutions to the transformer condition assessment problem, using the ER approach, are then illustrated highlighting the potential of the ER algorithm. The details of the computational steps of the application studies are discussed following the original ER algorithm. Based upon the outputs of the ER approach, system operators can obtain overall evaluations of observed units, as well as unit rankings. It can be deduced from the results that the ER approach is a suitable solution for combining multi-attribute information for transformer diagnosis purposes.

The second part of this chapter presents a hybrid diagnosis approach to the analysis of DGA data based upon several traditional DGA methods. The results demonstrate that the pseudo fuzzy representations of the three traditional DGA methods perform adequately over a wide range of test values taken from actual failed transformers, and that the hybrid system can effectively combine the evidence to produce a more meaningful and accurate diagnosis. The test and result section shows clearly the power of the ER algorithm to combine effectively all of the available evidence from three DGA diagnosis methods and provide an array of possible faults, mimicking the logical reasoning process of a DGA reviewer. It also demonstrates the practicality of using fuzzy membership functions for generating subjective beliefs in a simple manner using only two mathematical functions. The potential of this system lies in the fact that whereas other systems treat the problem as one of classification, ER treats the problem as one of reasoning based upon DGA data. The flexibility of the tree structure used to make decisions and the algorithm for combining the evidence means that the system can be extended easily to encompass new diagnosis techniques by simply adding extra branches parallel to the ones currently used.

The final part of this chapter describes a BN approach to transformer DGA interpretations, which is easy to construct and able to interpret with formal probabilistic semantics. The effectiveness of the IEEE and IEC coding schemes has been validated using the proposed BN approach, which is also able to handle the missing codes in the traditional DGA coding scheme. A BN has been created to diagnose transformer faults based upon the IEEE/IEC DGA ratio method. An applicable solution to a transformer DGA problem, using the BN approach, is illustrated to highlight the potential of BNs. It can be seen from the results that the proposed approach is capable of tackling the DGA problem for power transformers as a supportive tool along with the IEEE and IEC DGA coding scheme.

References

1. Mollmann A, Pahlavanpour B (1999) New guidelines for interpretation of dissolved gas analysis in oil-filled transformers. Electra, CIGRE France 186:30–51
2. International Electrotechnical Commission (1993) IEC600762-power transformers—part 2: temperature rise. International Electrotechnical Commission Standard, Geneva
3. International Electrotechnical Commission (1978) IEC60559: interpretation of the analysis of gases in transformers and other oil-filled electrical equipment in service. International Electrotechnical Commission Standard, Geneva
4. The Institute of Electrical and Electronics Engineers (1994) Transformers Committee of the IEEE Power Engineering Society, IEEE guide for the interpretation of gases generated in oil immersed transformers, IEEE Std. C57.104-1991. The Institute of Electrical and Electronics Engineers, Inc., New York
5. Saaty TL, Vargas LG (1988) The analytic hierarchy process, RWS Publications, Pittsburgh, pp 1–24
6. Rogers RR (1978) IEEE and IEC codes to interpret incipient faults in transformers using gas in oil analysis. IEEE Trans Electr Insul 13(5):348–354
7. Yang JB, Singh MG (1994) An evidential reasoning approach for multiple attribute decision making with uncertainty. IEEE Trans Syst Man Cybern 24(1):1–18
8. Yang JB, Xu DL (2002) On the evidential reasoning algorithm for multiple attribute decision making under uncertainty. IEEE Trans Syst Man Cybern A Syst Hum 32(3):289–304
9. Lin CE, Ling JM, Huang CL (1993) An expert system for transformer fault diagnosis and maintenance using dissolved gas analysis. IEEE Trans Power Deliv 8(1):231–238
10. Zaman MR (1998) Experimental testing of the artificial neural network based protection of power transformers. IEEE Trans Power Deliv 13(2):510–517

Chapter 9
Winding Frequency Response Analysis for Power Transformers

Abstract Nowadays, FRA has received great attention for transformer winding condition assessment, which has gradually replaced the conventional LVI method. It provides higher sensitivity to changes occurring in a transformer winding compared with the LVI method and can be performed relatively easily on-site. In this chapter, first the origin of FRA and the definition of FRA transfer functions are introduced. Two FRA methods, i.e. the impulse response method and the SFRA method, are presented alongside a typical experimental test setup for FRA. Then a brief literature review of winding models is given, which can be used for analytical calculations and circuit simulations for windings. The rest of this chapter focusses on FRA comparison techniques and the interpretation of frequency response measurements regarding various types of winding faults.

9.1 Introduction

Power transformers are specified to withstand mechanical forces arising from both shipping and subsequent in-service events, such as faults and lightening. Once a transformer is damaged, replacement cost of a large transformer may reach up to several million pounds in the U.K. If an incipient fault of a transformer is detected before it leads to a catastrophic failure, the transformer may be repaired on-site or replaced according to a scheduled arrangement. Therefore, conditions of critical transformers should be closely and continuously monitored in order to ensure maximum uptime. As a result, the so-called condition-based maintenance can reduce risks of forced outages and damages to adjacent equipment of a transformer.

Transformer in-service interruptions and failures usually result from dielectric breakdown, winding distortion caused by short-circuit withstand, winding and

W. H. Tang and Q. H. Wu, *Condition Monitoring and Assessment of Power Transformers Using Computational Intelligence*, Power Systems,
DOI: 10.1007/978-0-85729-052-6_9,

Fig. 9.1 A winding deformation fault found in a scrapped transmission transformer

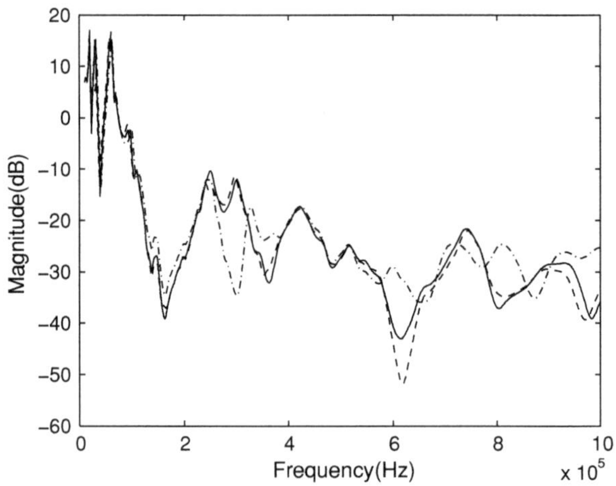

Fig. 9.2 FRA traces of a three-phase power transformer

magnetic circuit hot spot, failure of accessories such as OLTCs, bushings, etc. Winding distortion faults may cause catastrophic failures of transformers such as dielectric breakdown and short-circuits. A winding deformation fault found in a scrapped transmission transformer is shown in Fig. 9.1. The research into the field of winding deformation identification has drawn a great deal of attention over the past two decades.

The FRA test, first proposed by Dick and Erven [1], is a very sensitive technique for detecting winding movement faults caused by loss of clamping pressure or by short-circuit forces. In Fig. 9.2, three frequency response curves of a three-phase transformer are displayed in a wide range of frequency up to 1 MHz, each of which represents the FRA samples of one phase of the transformer. Due to different flux paths for the middle and side phases of the transformer, the FRA curves of two phases are very similar, which are different from that of the other phase as observed in Fig. 9.2. Variations in winding frequency responses may reveal a physical change inside a transformer, e.g. winding movement caused by loosened clamping structures or winding deformation due to shorted turns. In industrial practice, FRA is one of the most suitable diagnostic tools that can give an indication of winding displacement and deformation faults. It can be applied as a non-intrusive technique to avoid interruptive and expensive operations of opening a transformer tank and performing oil degasification and dehydration, which can minimise the impact on power system operations and loss of supply to customers and consequently save millions of pounds in timely maintenance. Most utility companies own databases containing historical FRA data for large power transformers. For example, NG regularly tests large transformers using SFRA in a frequency range up to 10 MHz. By comparing a frequency response measured during maintenance with a fingerprint measurement obtained at an earlier stage, FRA is widely employed by utility companies as a comparative method in the low frequency range of several tens of KHz to 1 MHz. Differences may reveal internal damages of a transformer, hence inspections can be scheduled for repairing. However, such a comparative method cannot quantify the change caused by a fault and reveal the location of the fault. It is, therefore, necessary to develop an accurate FRA modelling and reliable fault diagnosis approach to interpreting the physical meaning underneath variations of FRA data, which is with considerable industrial interest. Furthermore, a diagnosis framework dedicated to FRA is desirable in order to unify decision-making processes, when incomplete and imprecise data are involved.

9.2 Transformer Transfer Function

FRA is generally applied to a complex network of passive elements. A transformer is considered to be a complex network of RLC components, which is represented by three types of elements, i.e. resistors, inductors and capacitors. Such a distributed network contains an infinite number of small RLC components. The three elements are used to represent the resistance of copper windings, inductance of winding coils and capacitance from the insulation layers between coils, between a winding and a core, between a core and a tank, between a tank and a winding, etc. The main objective of FRA is to determine how the impedance of a test transformer behaves over a range of applied frequencies. The reactive properties of a test transformer are dependent upon and sensitive to changes in frequency. Most transformers produce a very distinct resonance in a specified frequency range. In the case of transformers, transfer functions are used for winding modelling and transient studies.

The transfer function of an RLC network is the ratio of the output and input frequency responses, when the initial conditions of such a network are zero. The phase relationships and magnitude can be extracted from the outputs of the transfer function. The idea of FRA for transformer winding condition assessment is based on the fact that winding displacement or deformation changes the geometrical properties of a winding, which are related to its internal capacitive and inductive parameters. The changes of these parameters consequently alter winding frequency responses, which can be observed by measuring the transfer function of a winding [2–6]. In an FRA test, this is the ratio k of the output voltage U_{out} over the input voltage U_{in}, represented as the logarithmic magnitude–frequency responses in dB–Hz as below:

$$k = 20\lg\left|\frac{U_{out}}{U_{in}}\right|, \tag{9.1}$$

and φ is the phase difference between U_{in} and U_{out}, which represents the phase–frequency response in degrees–Hz as the following:

$$\varphi = \angle\left(\frac{U_{out}}{U_{in}}\right). \tag{9.2}$$

9.3 Frequency Response Analysis Methods

There are two FRA test techniques, which inject test signals with a wide range of frequencies into a specimen. The first one is to inject an impulse into a winding and the second technique makes a frequency sweep using a sinusoidal signal source. The former is known as the LVI method and the latter as the SFRA method.

9.3.1 Low Voltage Impulse

When using the LVI method, an impulse voltage signal is injected into one terminal of a winding. The voltage at another terminal or the current passing through the winding connecting to the terminal or any of the other windings is measured. The signals are filtered and sampled in the time domain, which are converted into the frequency domain using the fast Fourier transformation (FFT) to extract information at individual frequencies. Finally, a transfer function is derived representing the FFT outputs at each frequency point of the measured signals. For example, an impulse can be applied to the HV terminal and the current at the HV neutral and the voltage transferred to the LV line terminal can be measured. Two types of transfer functions can be derived. One is between the HV current and the applied voltage and the other between the transferred voltage and the applied voltage.

The advantage of the LVI method is that it is possible to measure several currents and voltages simultaneously, which can reduce the time of each FRA test.

However, there are some disadvantages of this method, e.g. noise corruption during a test, poor resolution at low frequencies and limitation of excitation source in terms of energy.

9.3.2 *Sweep Frequency Response Analysis*

The SFRA method can obtain a measurement at each frequency point of interest by injecting a sinusoidal waveform at a constant magnitude. Then, the magnitude and phase shift measurements are sampled at predefined frequency points, which means it is a direct method to get frequency responses without using FFT over a specified frequency range. However, the SFRA method takes longer to produce a complete set of measurements compared with the impulse response method. A network analyser is normally used in industry for an FRA test. Such a test requires a 3-lead approach, with the leads providing signal, reference and test. In Fig. 9.3, a typical test connection is shown, which is with the 3-lead approach using a network analyser. The signal put into a winding is measured to provide a reference, which is then compared with the signal which emerges at the end of the winding and is measured by the test lead. This configuration can reduce the effect of test cables on test results. Each test lead comes with a cable shield ground, which is connected to the transformer at the base of test bushing to provide a common ground.

To sum up, the SFRA method provides a high signal-to-noise ratio using filtering to remove broadband noise. On the other hand, the time required to produce a frequency sweep depends on the frequency resolution and is usually longer than

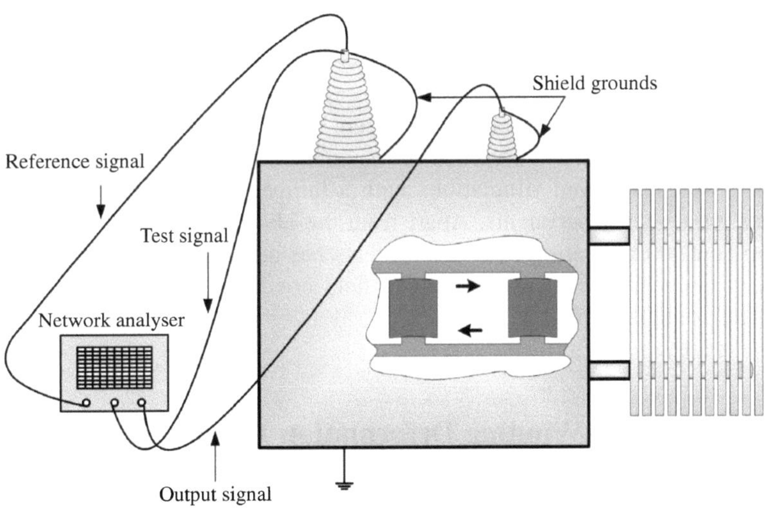

Fig. 9.3 Typical FRA test connection

that of the impulse response method. Currently, the utility industry prefers to use the SFRA method, which takes place automatically with a network analyser test set.

9.4 Winding Models Used for Frequency Response Analysis

A wide range of research activities have been undertaken to utilise and interpret FRA data for winding fault diagnosis, mainly including the development of accurate winding models and elaboration of FRA measuring systems. In [7] analytical expressions were used to estimate parameters of a lumped-parameter model based upon the geometry of a transformer. However, such a lumped-parameter model is limited in accuracy at the high frequencies from 1 to 10 MHz. Combinations of transfer functions [8] were employed to interpret the evolution of frequency responses, which are lack of physical representation regarding transformer structures. On the other hand, the theories of distributed parameter systems and travelling wave in transmission lines offer an appropriate mathematical foundation to model the propagation process of a voltage signal injected into a transformer winding. Rudenberg [9] elaborated and extended the travelling wave theory for lossless transformer winding analysis. In [10], each turn of a winding is represented as a single transmission line, which makes multi-conductor transmission line models complex to operate in case of analysis of a winding with a large number of turns. A high frequency power transformer model based on the finite element method (FEM) was applied in [11] for the accurate estimation of winding parameters for voltage stress calculations, which suggested that there was useful FRA information in the high frequency range. This FEM-based transformer model showed a high degree of accuracy compared with experiment FRA measurements and indicated that high frequency tests were capable of detecting small winding changes. However, the accurate simulation of high frequency behaviour of winding above 1 MHz can only be achieved with small sectioning of the above FEM model, which leads to its essential complexity with large computation time and complex FEM implementations. A recent study in [12] also demonstrated the potential for FRA result interpretation in an extend range of frequency up to 10 MHz, which involved simulations with a lumped-parameter model and comparisons with field experiments. Apart from the above modelling techniques, the development of elaborating FRA test systems has also attracted many researchers to obtain more precise measurements in field conditions by selecting an appropriate test configuration.

9.5 Transformer Winding Deformation Diagnosis

Amongst various diagnostic techniques applied to power transformer condition monitoring, only FRA is the most suitable diagnostic tool employed for reliable winding displacement and deformation assessment [13]. In practice, transformer winding condition assessment is conducted manually by experts or trained on-site engineers. They mainly tend to use a simple interpretation of measured frequency responses for transformer winding assessment incorporating "time-based", "construction-based" and "type-based" comparisons of FRA traces [14]. Despite extensive research on FRA result interpretations [13, 15], the decision-making procedure with FRA has not yet been formalised. The final diagnosis on the condition of a transformer winding is mainly made in a subjective manner depending on expert's experience [2, 6]. Often the diagnosis is not conclusive since different experts may give inconsistent judgements with regard to the same transformer. If it is the case, a transformer winding is inspected directly when removed out of its tank. Therefore, the effective and consistent combination of the available transformer diagnoses with the purpose to obtain a balanced overall condition evaluation is much desired based on collective expertise.

An FRA assessment process is mostly intuitive, since experts have to consider all available measurement data using cross-comparison techniques and then make a decision. Therefore, a reasonable idea is to integrate all the information derived from the available interpretation techniques using a formalised and meaningful assessment framework, which can produce a balanced overall evaluation. In another word, decisions made on the basis of each FRA result comparison technique can be considered as a piece of evidence for the overall assessment of winding conditions. In this respect, a condition assessment process can be regarded as an MADM problem. Since subjective diagnoses given by experts are sometimes imprecise and even incomplete, an ER approach based upon the Dempster–Shafer theory [16, 17] is utilised in this research for evidence aggregation to integrate expert's judgements involving uncertainties.

As mentioned above, changes in resonance frequencies or/and magnitudes are linked to deviations of inductances or capacitances, which are defined by physical dimensions of a transformer. Thus, changes in resonances are, in fact, the evidence used for diagnosis of winding mechanical faults, such as displacement and deformation. The analysis of frequency response traces was, at the first, attempted by Dick and Erven [1], who explained winding frequency domain behaviours by introducing the terms of low, medium and high frequency ranges. Below the 10–20 kHz bound a transformer winding response is dominated by inductive components, whereas in the medium frequencies from 10–20 kHz to 1 MHz the combinations of inductances and capacitances cause multiple resonances over the frequency range. Further raise of frequency leads to the case when distributive capacitances of a winding tend to shunt winding inductances and resistances in the 2–10 MHz frequency range. Another frequency bandwidth division was proposed on the basis of experimental case studies, where the relations between different frequency ranges of an FRA trace and RLC elements of a transformer were analysed [2, 18]. According to [18, 19], FRA traces can be interpreted based upon separating frequency responses into four frequency bands being more sensitive to different winding faults as indicated in Table 9.1, which is adopted in this research.

Table 9.1 Sub-band division of frequency responses [18]

Frequency	Failure sensitivity
<2 kHz	Core deformation, open-circuits, shorted turns and residual magnetisation
2–20 kHz	Bulk winding movement between windings and clamping structure
20–400 kHz	Deformation within the main or tap windings
400 kHz–about 1 MHz	Movement of the main and tap windings, ground impedance variations

9.5.1 Comparison Techniques

Three comparison techniques are usually practiced in industry for FRA result interpretation and transformer winding condition assessment, which are introduced as below.

The "time-based" (reference) comparison (RC) is considered as the most effective technique for detection of changes in a transformer winding [2]. It includes a comparison of an FRA trace in hand with a reference response of a transformer winding, being taken when the transformer is known to be in a normal condition.

The comparison between FRA results and the corresponding data taken from other legs in separate tests for a multi-legged transformer is called a "construction-based" (phase) comparison (PC). This procedure can be implemented only for certain types of transformers having a "star" phase connection. Owing to the difference of flux paths for the middle and side phases of a transformer, there are expected differences between low frequency responses during an open-circuit test. A short-circuit test [5] is used widely to eliminate the core effect, thereby allowing the direct comparison of the responses between different phases over low frequencies.

The third procedure, namely the "type-based" (sister unit) comparison (SUC), utilises FRA measurements taken from identically designed transformers produced from the same manufacturer [14]. If poor phase comparisons are shown between an FRA trace and the trace of the same phase winding of the same type transformer with respect to resonance frequencies and magnitudes in low and medium frequency ranges, it may indicate potential internal changes of the investigated transformer winding, knowing the sister transformer winding is normal.

9.5.2 Interpretation of Frequency Response Measurements

In general, a visual cross-comparison of frequency responses aims to detect newly appeared suspicious deviations of an FRA trace compared with various etalon responses. The appearance of clear shifts in resonance frequencies or new resonant points may characterise faulty conditions of a winding. Although FRA is able to detect failure presence within a winding, the classification of different winding failure modes still demands more research in order to establish standardised criteria. Nevertheless, some general criteria regarding FRA interpretations on winding conditions can still be extracted, which are summarised below. It should be noted that although these criteria are rooted on experimental case studies, reported in various FRA research publications, and frequency response simulations performed by the authors, they are not conclusive to establish unambiguous decisions in some occasions.

9.5.2.1 Normal Winding (NW)

The normal state of a winding usually corresponds to a consistent response shape and resemblance between several responses at cross-comparison. However, observed in most cases small variations, which may appear due to phase differences at the "construction-based" (phase) comparison, can be disregarded considering a corresponding response of a sister transformer.

For instance, consider a 90/33 kV 75 MVA transformer, frequency responses of which are presented in the form of the transfer function in Eq. 9.2 as shown in Fig. 9.4. The transformer is in a normal condition after a health inspection. As seen from the figure, there is a clear repeatability of the trends corresponding to different phases of the transformer in the low and medium frequency ranges up to 600–700 kHz and a slight deviation between phases at higher frequencies.

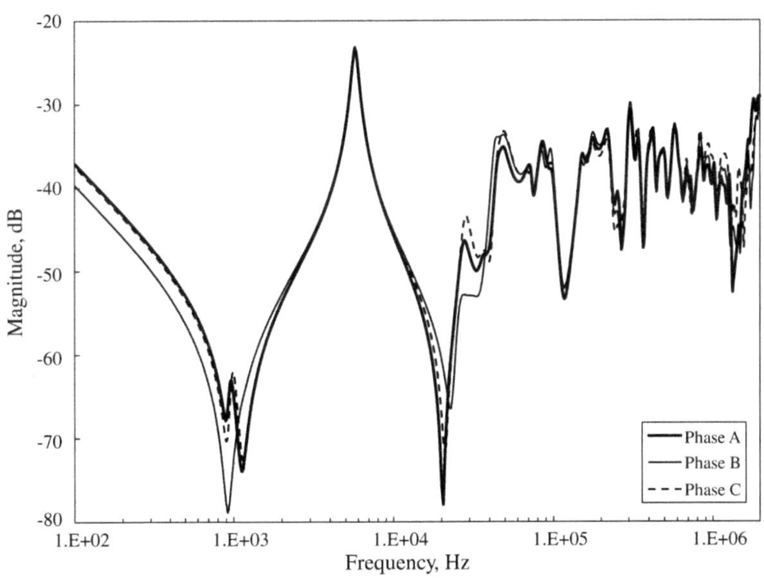

Fig. 9.4 "Construction-based" (phase) comparison (HV side, log scale)

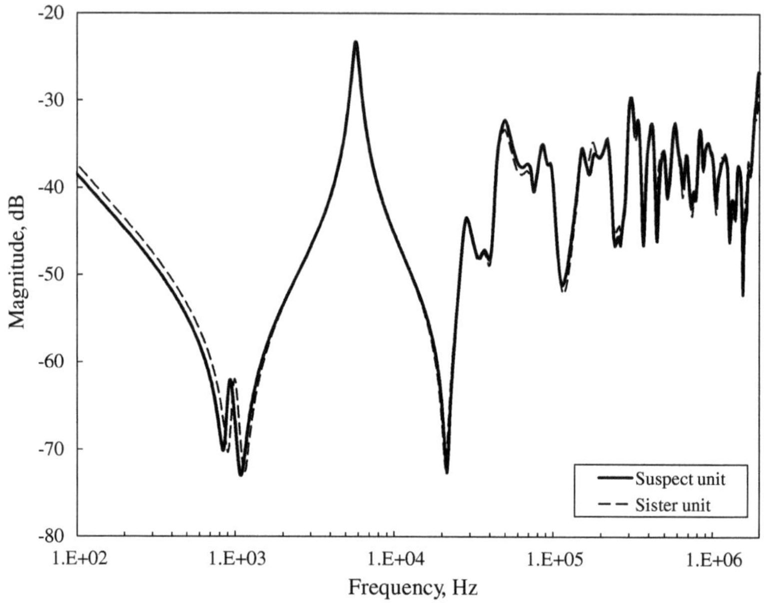

Fig. 9.5 "Type-based" (sister unit) comparison (HV side, phase C, log scale)

Particularly, the phase C response shows the clearest deviations at the last several resonance points with respect to the other two phases. The concern about the condition of the phase C winding can be eliminated by comparing two responses taken from the C phases of the investigated and a normal sister transformer of the same type due to a good resemblance of the responses, as illustrated in Fig. 9.5. This is discussed in detail in Sect. 11.4.1 where an example describing the condition evaluation of transformer windings is given.

9.5.2.2 Short-Circuited Turns (SCT)

Experience shows that SCT in a winding can be identified clearly due to distinguishable disappearances of the first resonance points at low frequencies associated with a transformer core, which are also seen at short-circuit test FRA measurements [5]. Therefore, this fault can be detected easily by using only the "construction-based" (phase) comparison without additional expert analysis or a comparison with reference responses [20].

9.5.2.3 Clamping Failure (CF)

A CF is rare and may be caused by a bulk winding movement. Available case studies show clear shifts to the right in low frequency resonances below 20 kHz whilst there are no significant variations at higher frequencies compared with sister unit phase responses [21].

9.5.2.4 Axial Displacement (AD)

Right shifts in medium frequency resonances may indicate potential AD, also known as axial collapse, of a winding with respect to other transformer windings. In addition, new resonances may appear at higher frequencies [2, 18, 19, 22]. Similar to a hoop buckling (HB), this failure usually attributes no clear indication at low frequencies.

9.5.2.5 Hoop Buckling

Severe radial deformation of a winding, known as a HB, leads to a bent winding, being not broken. These deformation faults normally occur in inner (usually LV) windings and show significant decrease (shift to left) of the medium frequency resonance points whilst low frequency open-circuit responses usually indicate no difference [2, 7]. In addition, short-circuit test results may reveal an increased input impedance of the damaged phase with respect to other phases of a transformer [19, 23].

9.5.2.6 Residual Magnetisation (RM)

As stated in Table 9.1, RM changes a response in the very low frequency region of open-circuit test measurements of a transformer. This is mainly shown in consistent magnitude deviations with possible slight shifts of low frequency resonances. In contrast, the overall response shape remains consistent compared with the reference ones [2, 24]. RM does not normally affect responses above 20 kHz, which can be easily eliminated using core demagnetisation techniques.

9.5.2.7 Poor Grounding (PG)

The connection to ground of measurement cables is very important for obtaining repeatable and reliable FRA measurements due to its considerable effect on high frequency responses. Poor cable grounding is normally detected via clear unexpected magnitude deviations at higher frequencies compared with other phase measurements [2, 5]. This could be a tremendous source for misinterpretation of FRA results [25]. Poor tank grounding is easier to spot since it affects frequency response measurements of all windings of a transformer [2, 3, 26]. Although, the last two diagnosis cases, i.e. RM and PG, only affect FRA measurements, and do not indicate any change in a winding, their presence may lead to misinterpretation of FRA results. Therefore, these diagnoses need to be taken into account during an FRA assessment.

9.6 Summary

FRA is one of the most sensitive tests for winding deformation and displacement faults. The existing FRA tests include the SFRA test and the LVI test. Three trace comparison techniques are employed widely for FRA result interpretation, which require trained experts to interpret measured FRA traces in a subjective manner. Based upon the trace comparison techniques, difference between baseline traces and newly measured traces may indicate a winding fault. A number of general criteria regarding FRA interpretations have been summarised in this chapter, including NW, SCT, CF, AD, HB, RM and PG. These fault types will be used in a later chapter for evidence-based winding condition assessment. In addition, several well-known winding models are introduced briefly, which can provide accurate analytical calculations and circuit simulations to represent transformer windings.

References

1. Dick EP, Erven CC (1978) Transformer diagnostic testing by frequency response analysis. IEEE Trans Power Appar Syst PAS-97(6):2144–2150
2. Ryder SA (2003) Diagnosing transformer faults using frequency response analysis. IEEE Electr Insul Mag 19(2):16–22
3. Ryder SA (2002) Transformer diagnosis using frequency response analysis: results from fault simulations. In: 2002 IEEE Power Engineering Society summer meeting, vol 1, pp 399–404
4. Mcgrail T (2003) SFRA basic analysis, vol 1, version 1.0. Doble Engineering Co.
5. Mcgrail T (2003) SFRA basic analysis, vol 2, version 1.0. Doble Engineering Co.
6. Wilson A, McGrail T (2002) The need for and use of techniques to assess the mechanical integrity of transformer windings. In: Proceedings of the 2002 international conference of Doble clients, 10 pp
7. Rahimpour E et al (2003) Transfer function method to diagnose axial displacement and radial deformation of transformer windings. IEEE Trans Power Deliv 18(2):493–505
8. Pleite J et al (2002) Transformer modeling for FRA techniques. In: Proceedings of the IEEE Power Engineering Society transmission and distribution conference, ASIA PACIFIC, vol 1, pp 317–321
9. Rudenberg R (1968) Electrical shock waves in power systems: traveling waves in lumped and distributed circuit elements. Harvard University Press, Cambridge
10. Shibuya Y, Fujita S, Hosokawa N (1997) Analysis of very fast transient overvoltage in transformer winding. IEE Proc Gener Transm Distrib 144(5):461–468
11. Bjerkan E, Høidalen H (2007) High frequency FEM-based power transformer modeling: investigation of internal stresses due to network-initiated overvoltages. Electric Power Syst Res 77:1483–1489
12. Wang M, Vandermaar A, Srivastava KD (2005) Improved detection of power transformer winding movement by extending the FRA high frequency range. IEEE Trans Power Deliv 20(3):1930–1938

13. Wang MG (1994) Winding movement and condition monitoring of power transformers in service. The University of British Columbia, Department of Electrical and Computer Engineering, Vancouver
14. Christian J, Feser K (2004) Procedures for detecting winding displacements in power transformers by the transfer function method. IEEE Trans Power Deliv 19(1):214–220
15. Jeong SC, Kim JW, Park P, Kim SW (2005) A pattern-based fault classification algorithm for distribution transformers. IEEE Trans Power Deliv 30(4):2483–2492
16. Yang JB, Sen P (1994) A general multi-level evaluation process for hybrid MADM with uncertainty. IEEE Trans Syst Man Cybern 24(10):1458–1473
17. Yang JB, Xu DL (2002) On the evidential reasoning algorithm for multiple attribute decision making under uncertainty. IEEE Trans Syst Man Cybern 32(3):289–304
18. Kennedy GM, McGrail AJ, Lapworth JA (2007) Transformer sweep frequency response analysis (SFRA). Energize, eepublishers, pp 28–33
19. Prout P, Lawrence M, McGrail T, Sweetser C (2004) Substation diagnostics with SFRA: transformers, line traps and synchronous compensators. In: Proceedings of the 2004 EPRI substation diagnostics conference, 22 pp
20. Sweetser C, McGrail T (2005) Field experiences with SFRA. In: Proceedings of the 2005 international conference of Doble clients, 11 pp
21. Kennedy GM, Sweetser C, McGrail T (2006) Field experiences with SFR. In: Proceedings of the 2006 international conference of Doble clients, 8 pp
22. Nirgude PM, Channakeshava BG, Rajkumar AD, Singh BP (2005) Investigations on axial displacement of transformer winding by frequency response technique. In: Proceedings of the XIVth international symposium on high voltage engineering, Beijing, China, 2005, 4 pp
23. Prout P, Lawrence M, McGrail T, Sweetser C (2004) Investigation of two 28 MVA mobile transformers using Sweep Frequency Response Analysis (SFRA). In: Proceedings of the 2004 international conference of Doble clients, 2004, 13 pp
24. Abeywickrama N, Serdyuk YV, Gubanski SM (2008) Effect of core magnetization on frequency response analysis (FRA) of power transformers. IEEE Trans Power Deliv 23(3):1432–1438
25. Homagk C, Leibfried T, Mössner K, Fischer R (2007) Circuit design for reproducible on-site measurements of transfer function on large power transformers using SFRA method. In: 15th International symposium on high voltage engineering, Slovenia, 2007
26. Ryder SA (2001) Frequency response analysis for diagnostic testing of power transformers. Electr Today 13(6):14–19

Chapter 10
Winding Parameter Identification Using an Improved Particle Swarm Optimiser

Abstract Among various transformer winding models, the lumped-element model is the most commonly used, as it gives a satisfactory representation of a real transformer. This chapter is concerned with a model-based approach to identifying distributed parameters of the lumped-element winding model using an improved particle swarm optimiser (PSO) technique. A simplified circuit of the lumped-element model is developed to calculate frequency responses of transformer windings in a wide range of frequency domain. In order to seek optimal parameters of the simplified winding circuit, a particle swarm optimiser with passive congregation is employed to identify model parameters based on frequency response samples. Simulations and discussions are presented to investigate the potential of the proposed approach.

10.1 Introduction

As discussed in Chap. 9, in practice, FRA (frequency response analysis) is usually considered as a comparative technique. Using statistical analysis methods, an FRA trace is compared with an FRA baseline or a selected FRA reference trace, and poor FRA comparisons may indicate a winding fault. For instance, a root mean square (RMS) method can be employed to calculate the difference between two frequency response curves, i.e. a reference trace and a trace of interest. A set of factors derived from the RMS method can provide indications of winding faults. However, it is difficult to quantify a fault severity level and further to locate a fault according to the RMS factors. In this chapter, a model-based approach using an evolutionary algorithm is developed to tackle the above problems. This approach treats an FRA interpretation process as a model-based evaluation process, the main problem of which is to identify accurately winding model parameters to detect a

W. H. Tang and Q. H. Wu, *Condition Monitoring and Assessment of Power Transformers Using Computational Intelligence*, Power Systems,
DOI: 10.1007/978-0-85729-052-6_10,

fault and predict failure tendencies for a transformer. It is considered in this study that, if variations of distributed parameters of a transformer can be quantified, a diagnostic procedure can then be implemented based on the assessment of the identified parameters for fault diagnosis purposes.

In this chapter, an improved particle swarm optimiser (PSO), namely PSOPC (particle swarm optimiser with passive congregation), is employed to identify the parameters of a ladder network model for representing a power transformer winding. With the use of PSOPC, the distributed parameters of the ladder network model can be identified based on FRA measurements with a faster convergence rate. The derived distributed parameters can be further utilised to detect winding deformation faults. Simulation results and discussions are addressed at the end of this chapter.

10.2 A Ladder Network Model for Frequency Response Analysis

It is well known that there is a direct relationship between the geometric configuration of a winding and a core in a transformer and the distributed network of resistances, inductances and capacitances that make it up. In a wide range of frequency domain (2 kHz $< f <$ 2 MHz, where f denotes frequency), a transformer winding behaves as a complex ladder type network consisting of a series of inductances, capacitances, resistances and conductances. For a transformer winding with n sections, a simplified equivalent circuit, which was first proposed in [1], is shown in Fig. 10.1, where L_n denotes the winding inductance per section, C_n the ground capacitance per section, C_b the busing capacitance, K_n the series capacitance per section, R_i the input impedance, R_o the output impedance and V_s the input voltage source. The distributed parameters of a ladder RLC network can be determined through experiments or based on its frequency-dependent responses.

In industrial practice, an FRA test measures frequency responses of a winding at each frequency point of interest. An excitation voltage source, i.e. V_s, generates a sinusoidal waveform at a constant magnitude, which is applied to the test terminals of a winding. Since the source magnitude is constant and can be maintained for a specified amount of time, designated digitisers have ample time to adjust their gain settings, resulting in higher dynamic range performance. As mentioned above,

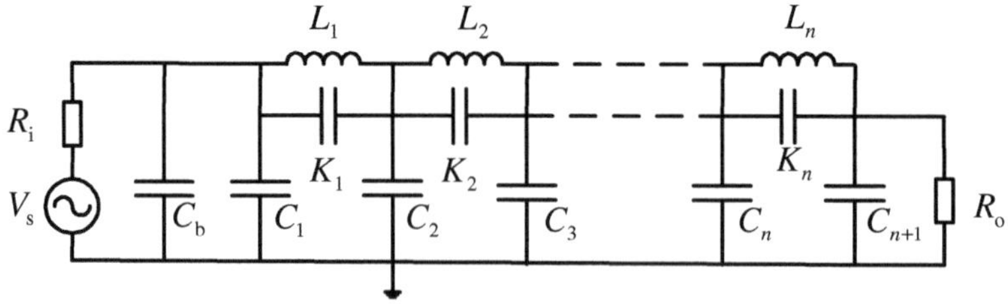

Fig. 10.1 A ladder network model for representing a transformer winding

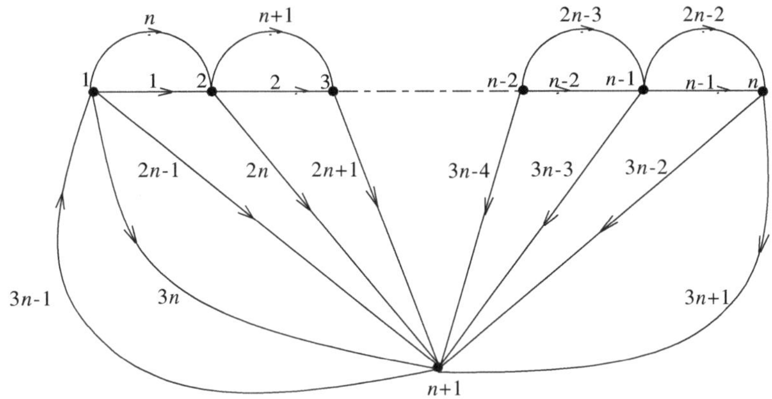

Fig. 10.2 Circuit nodal graph of a ladder network model

FRA is applied to the complex network of passive elements, which is depicted in Fig. 10.1. In order to calculate frequency responses of the lumped-element circuit, the ladder network model is transferred into a circuit nodal graph as shown in Fig. 10.2. The derived nodal graph contains $n + 1$ nodes and m branches ($m = 3n + 1$), which is mapped directly to a winding with n sections with respect to the RLC circuit illustrated in Fig. 10.1.

By applying circuit nodal analysis principles, the voltage of each node in Fig. 10.1 is calculated with the following equation:

$$U = G^{-1}I \tag{10.1}$$

where U is the node voltage vector, **G** the node conductance matrix and I the excitation current vector, i.e. $I = [-V_s/R_i, 0, \ldots, 0]^T$. The structure and the number of winding sections of the RLC circuit model can be varied to mimic a real transformer through programming. It is obvious that such a model can be used for transformer FRA simulations, distortion location analysis and other relevant research regarding transformer windings. On the basis of Eq. 10.1, the established ladder network model is employed to simulate frequency responses of a transformer winding in the next section.

10.3 Model-Based Approach to Parameter Identification and Its Verification

10.3.1 Derivation of Winding Frequency Responses

The main goal of FRA is to measure the impedance $Z(j\omega)$ of a test transformer, where $j\omega$ denotes the presence of a frequency-dependent function and ω equals $2\pi f$. Since a FRA test uses a resistance R_o to match a measuring system, the output resistance R_o must be incorporated into the calculation of frequency response $H(j\omega)$. The relationship between $Z(j\omega)$ and $H(j\omega)$ is shown in Eq. 10.2:

$$H(j\omega) = \frac{U_{out}}{U_{in}} = \frac{R_o}{Z(j\omega) + R_o}. \tag{10.2}$$

The preferred method for illustrating $H(j\omega)$ graphically is to use a Bode diagram, which plots the magnitude as $20\log_{10}(H(j\omega))$. Therefore, the unit of the derived frequency responses is decibel (dB), which is the output of the lumped-element model for model optimisation discussed in the next subsection.

As known, the distortion level of a winding distortion fault should be carefully assessed before removing a transformer from service to avoid an unnecessary outage. At present, there is not an effective method to scale and locate a winding distortion fault in the current industry practice. In the proposed model-based approach, it is considered that if prior-fault and post-fault parameters of a transformer model are known, the position and degree of a distortion fault may then be determined. Therefore, the proposed technique is employed to search optimal winding parameters by minimising the difference (i.e. fitness) between original frequency responses and simulated model outputs using PSOPC.

10.3.2 Fitness Function Used by PSOPC

The PSOPC is employed to identify model parameters to achieve the minimum fitness; hence, the outputs of the simplified circuit model have satisfactory agreements with the original frequency responses. Before implementing PSOPC and searching optimal parameters for the lumped-element model, a fitness function and other relevant parameters of a PSOPC programme should be defined. In this particular task, the errors between the original responses and the model outputs are defined as fitness. Thereby, for each individual (particle) of a population in PSOPC, its total fitness value is given as follows:

$$\min \sum_{i=1}^{N} \|u_{o_i} - a(\mathbf{G})^{-1} I\|, \tag{10.3}$$

where $u_{o_i} \in \mathbb{R}^{1\times n}$ is the original frequency response, $a \in \mathbb{R}^{1\times n}$ is a vector, i.e. $a = [0,...,0,1]$ and N is the number of original FRA samples involved for PSOPC optimisation.

10.4 Simulations and Discussions

In order to implement the proposed approach to determining the parameters of the lumped-element model with PSOPC learning, simulations and optimisations are implemented in MATLAB. Firstly, a simulated FRA test is carried out to generate frequency responses of the lumped-element model. The distributed parameters of the lumped-element model are predefined for simulation purposes. Then, the parameters of the lumped-element model are identified using PSOPC based on the simulated frequency responses. Results and discussions are addressed at the end of this section.

10.4.1 Test Simulations of Frequency Response Analysis

When a transformer is subjected to a real FRA test, the leads are configured in such a manner that four terminals are used. These four terminals can be divided into two unique pairs, one pair for the input and the other pair for the output. These terminals can be modelled in a two-terminal pair or a two-port network configuration. The following procedures are used to simulate a real FRA test:

1. According to the construction knowledge of a real transformer, the parameters of the lumped-element circuit are preset as $L_n = 10^5$ mH, $K_n = 100$ μF and $C_n = 100$ μF for each section of a winding. The number of winding sections is selected as 10 for an illustration purpose.
2. After the pre-definition of model parameters and its structure, a sinusoidal waveform at a constant magnitude 1.0 is applied to the simplified circuit as an input source at frequencies varying from 20 Hz to 1 MHz. Then, the magnitude of the generated frequency responses are recorded as data set *a* derived from Eqs. 10.1 and 10.2, which is employed as training targets for PSOPC optimisation.

10.4.2 Winding Parameter Identification

Based on the simulated data set *a*, PSOPC is utilised to identify winding parameters, which represent a lumped-element winding model with frequency responses close to data set *a*. The generation number of PSOPC is set as 50, and its fitness function is defined as Eq. (10.3). The following steps describe the PSOPC optimisation procedures:

1. To produce simulated frequency responses by feeding the frequency data and voltage source of data set *a* to the RLC circuit model with 10 sections representing a winding, the parameters of the circuit model are to be identified using PSOPC.
2. The optimisation procedures follow the PSOPC algorithm listed in Table 2.3.
3. When the optimisation termination criterion is reached, the model outputs and data set *a* are compared to verify the effectiveness of the proposed approach.

10.4.3 Results and Discussions

In Fig. 10.3, the original frequency response trend and the calculated frequency response trend using the parameters identified by PSOPC are displayed for a comparison. During optimisation, the fitness did not decrease after 50 generations, and there were still deviations between the two trends at the lowest peak point. The total fitness was 430 with respect to 400 sets of data. It is noted that the model output frequency response curve using the parameters identified by PSOPC learning is very close to the original responses of data set *a*. The identified parameters using PSOPC are $L_n = 0.939 \times 10^5$ mH, $K_n = 97.5$ μF and $C_n = 96.5$ μF, which are fairly close to the preset values. The results demonstrate that the proposed approach can find an accurate solution from a simulated FRA data set.

In comparison, an SGA is also utilised to identify the distributed parameters of the simulated winding model with data set *a* following the GA algorithm listed in Fig. 5.5. However, the obtained results are not satisfactory regarding the difference between data set *a* and the model outputs with GA learning. It is deduced that, PSOPC has a comparable or an even superior search performance for some hard optimisation problems with faster convergence rates [2], compared with other stochastic optimisation methods, such as SGA. In PSOPC, particles fly around in a multidimensional search space. During a flight, each particle adjusts its position according to its own experience and the experience of a neighbouring particle. The actions taken by each particle make use of the best position encountered by itself and its neighbour. Thus, PSOPC can combine local search methods with global search methods, which attempts to balance exploration and exploitation. It is summarised after a comparison study between the PSOPC learning and the SGA simulation in this research as follows:

1. An advantage of PSOPC is that GAs have at least four parameters, i.e. mutation probability, crossover probability, selection probability and maximum generations, to be tuned; in comparison, PSOPC has only two parameters to adjust, i.e. inertia weight and maximum generations, that makes it particularly attractive from a practitioner's point of view.

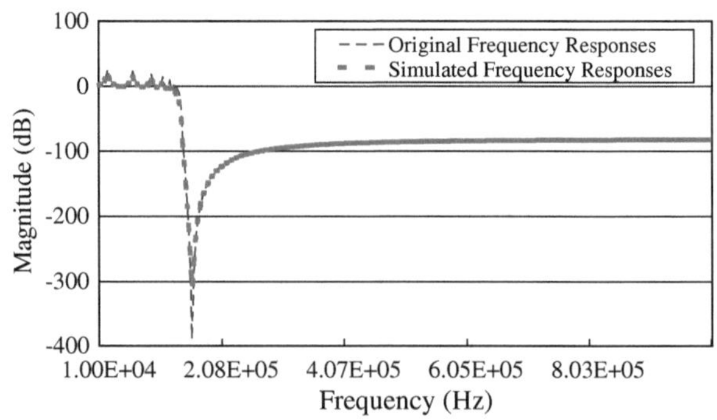

Fig. 10.3 Comparison between the target frequency responses (data set *a*) and the model outputs

2. Simulations have been carried out using both the SGA and the PSOPC for parameter identification, which employ the same fitness function and the same data extracted from data set a. With regard to a predefined fitness, PSOPC has shown a faster convergence rate than that of SGA in this particular task.

10.5 Summary

In summary, a model-based approach has been developed in this chapter to determine the parameters of a lumped-element winding model using PSOPC. A circuit nodal analysis technique has been applied to construct a generic model for transformers with variant winding sections. The PSOPC learning has delivered a satisfactory performance during optimisation based upon original FRA targets. Compared with other parameter estimation techniques, the proposed PSOPC has the advantages of a fewer parameters to adjust, a faster convergence rate and more local searches. There is a slight difference between the identified parameters and the preset parameters, which is negligible in a practical sense. It can also be deduced that the proposed approach is applicable and practical, which can be utilised for fault identification and trend analysis for detecting winding deformation faults. In addition, as the proposed approach has a simple form and a clear physical meaning, it holds significant potential for accurate condition assessment of transformer windings.

References

1. Dick EP, Erven CC (1978) Transformer diagnostic testing by frequency response analysis. In: IEEE transactions on power apparatus and systems, PAS-97, vol 6, pp 2144–2150
2. He S, Wu QH, Wen JY, Saunders JR, Paton RC (2004) A particle swarm optimizer with passive congregation. BioSystems 78(1–3):135–147

Chapter 11
Evidence-Based Winding Condition Assessment

Abstract A winding condition assessment process using FRA can be treated as an MADM problem by combining subjective and quantitative evidence. In this chapter, first an FRA assessment process is introduced briefly, which is then mapped into an MADM problem under an ER framework adopted for winding condition assessment. Subsequently, several examples of transformer winding condition assessment problems are presented using two ER evaluation analysis models, where the potential of the ER approach in combining evidence and dealing with uncertainties is demonstrated. In the case when more than one expert is involved in an FRA assessment process, the developed ER framework can be used to aggregate subjective judgements and produce an overall evaluation of the condition of a transformer winding in a formalised form.

11.1 Knowledge Transformation with Revised Evidential Reasoning Algorithm

As mentioned in Sect. 9.5, a winding condition assessment process using FRA can be considered as an evidence combination process, which depends largely on experts' experience. Usually, an expert makes a decision based upon all available information in hand and analyses it using cross-comparison techniques. However, there is no formalised framework established for implementing winding condition assessment, and uncertainties could be arising from a decision-making process. In addition, reference responses from the same phase of a transformer are usually unavailable in most cases [1], which increases uncertainties during assessment.

As discussed in Sect. 3.2, the ER approach provides a mathematical framework for combining uncertain information such as expert's subjective judgements.

W. H. Tang and Q. H. Wu, *Condition Monitoring and Assessment of Power Transformers Using Computational Intelligence,* Power Systems,
DOI: 10.1007/978-0-85729-052-6_11,

By considering each piece of information as a piece of evidence either supporting or denying a hypothesis, which corresponds to a winding fault type derived with FRA, the validity of all possible hypotheses can be calculated [2]. Considering a set of possible winding faults as listed in Sect. 9.5.2, a set of common hypotheses (evaluation grades) H is defined for an FRA assessment process with the revised ER as follows:

$$H = \{\mathrm{NW},\ \mathrm{SCT},\ \mathrm{RM},\ \mathrm{CF},\ \mathrm{HB},\ \mathrm{AD},\ \mathrm{PG}\}. \tag{11.1}$$

To describe a vague or imprecise information during fault diagnosis, an expert may use probabilities of a particular condition presence of a winding to make a decision, e.g. 50% chance that a transformer winding is *Normal* and 40% that the winding has suffered from or is currently with a *Hoop Buckling* failure. These probabilities are referred as degrees of belief and play an important role in an ER algorithm. Note that the above assessment example is incomplete as the total degree of belief is 50% + 40% = 90%, which is less than 100%. The missing 10% in such an assessment process represents the degree of uncertainty in expert's judgements. In this chapter, the revised ER algorithm is employed to deal with such uncertainties illustrated with two typical winding diagnosis examples.

11.2 A Basic Evaluation Analysis Model

Having defined a set of evaluation grades, a decision process of FRA assessment can be represented by a basic evaluation analysis model as shown in Fig. 11.1, which consists of three levels: general attributes Y, evaluation grades H and basic attributes E. Concerning FRA of transformer windings, only one general attribute y is considered. This is the overall evaluation of the condition of the investigated transformer winding (windings), which is supported by a set of basic attributes E comprising the subjective judgements derive from the diagnoses of RC (e_1), PC (e_2) and SUC (e_3), i.e. the reference, phase and sister unit comparisons, respectively, as introduced in Sect. 9.5.1.

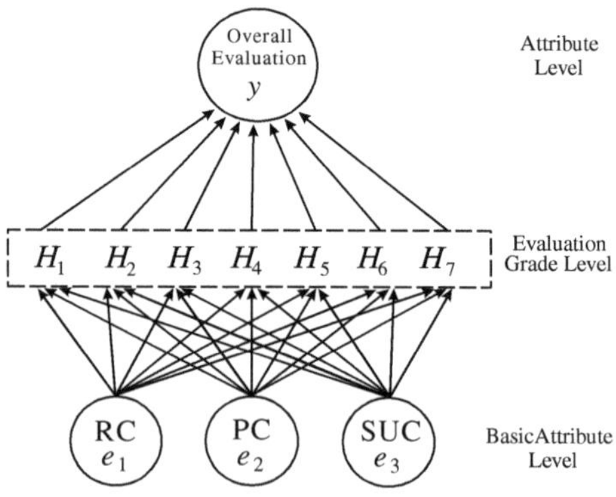

Fig. 11.1 A basic evaluation analysis model for transformer winding assessment using FRA

In the basic attribute level, an expert, with regard to available FRA data, formulates a unique evaluation for each of the attributes using the predefined H of evaluation grades. For instance, suppose that reference responses are unavailable at the time and on the basis of PC an expert suggests that there is a 50% possibility that a transformer has *Residual Magnetisation* of its core and a 50% belief that its winding is *Normal*. At the same time, the expert is 80% sure that the winding is *Normal* based on SUC. If these evaluations for each basic attribute are considered as the evidence of diagnostic information then, by combining them, an overall evaluation or diagnosis on the transformer winding condition can be made. In this way, a transformer winding condition assessment problem is represented in the form of a basic evaluation analysis model as shown in Fig. 11.1. The core of the model is the ER algorithm [3, 4], which is used to aggregate attributes in order to derive a balanced overall assessment, and the revised ER algorithm is employed in this chapter as introduced in Sect. 3.2.2.

11.3 A General Evaluation Analysis Model

Since there are no universally recognised quantitative criteria being established for an FRA assessment, more than one expert could be involved to provide an

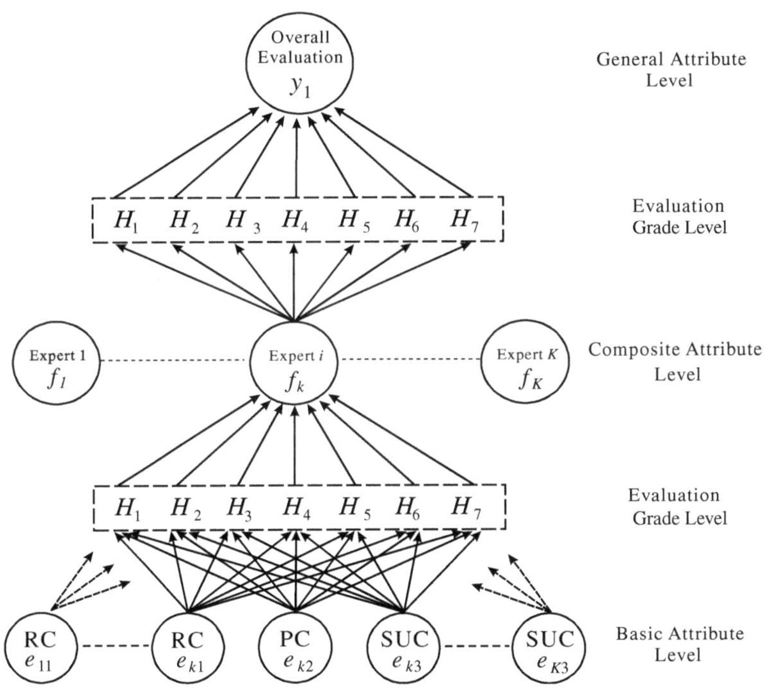

Fig. 11.2 A general evaluation analysis model for transformer winding assessment using FRA

independent evaluation using available frequency responses. In that case an overall evaluation has to be a result of combination of all the experts' judgements. Therefore, the assessment process can be regarded as a multilevel hierarchical analysis model, i.e. a general evaluation analysis model, which is shown in Fig. 11.2. The FRA assessment, thus, becomes a hierarchical analysis problem, where a high level attribute is assessed through associated lower level attributes. For instance, the overall diagnosis y_1 from K experts is obtained by combining the expert's subjective judgements f_k, ($k = 1,\ldots,K$), as in the composite attribute level. Each subjective judgement f_k is calculated via the individual consideration of the corresponding basic attributes e_{k1}, e_{k2} and e_{k3} in the form of the basic evaluation analysis model.

11.4 Results and Discussions

11.4.1 An Example Using the Basic Evaluation Analysis Model

Consider a simple example for evaluating the HV winding condition of a 90/33 kV 75 MVA power transformer using the revised ER algorithm considering phase and sister unit comparisons. No reference frequency responses are available in this example.

According to the basic evaluation analysis model defined in Sect. 11.2, a winding assessment problem is basically a combination of subjective decisions derived from three different types of comparisons, i.e. RC, PC and SUC comparisons, as introduced in Sect. 9.5.1. All the three techniques have to be assigned by normalised weights, which reflect their relative importance during an FRA assessment process.

In general, the reference comparison is considered as the most reliable method for FRA assessment [5, 6, 7]. However, when reference responses are not available, the phase response comparison or sister unit response comparison methods are employed instead. With regard to FRA result interpretations, Ryder [5] emphasised the slight advantage of PC over SUC. On the other hand, based on the extensive field experience Doble researchers suggest the sister unit comparison method [6] is a more beneficial technique. Since the relative importance ranking of these comparative techniques needs further investigations, both the methods are assumed to be of almost equal importance with a slightly higher importance of SUC in this example for illustrative purposes only.

Ranking the above three comparison methods, on the basis of Table 8.3 and the AHP method for weight determination given in Sect. 8.2.3, a general comparison matrix, in the form of (8.11), can be formed as below:

$$\mathbf{A} = \begin{bmatrix} 1/1 & 4/1 & 3/1 \\ 1/4 & 1/1 & 1/2 \\ 1/3 & 2/1 & 1/1 \end{bmatrix}, \qquad (11.2)$$

for attributes e_1, e_2 and e_3 of the basic evaluation analysis model shown in Fig. 11.1. Consequently, the corresponding relative weights are derived as $\omega_1 = 0.6232$, $\omega_2 = 0.1373$ and $\omega_3 = 0.2395$.

Consider Figs. 9.4 and 11.3, where the comparisons of phase responses of the investigated transformer are presented using the log and linear scales for more detailed analysis of the frequency responses at low and high frequencies, respectively. Overall, the log scale phase comparison in Fig. 9.4 gives no clear unexpected deviations between phases in the low and medium frequency diapasons. However, as pointed out in Sect. 9.5.2, one may observe clear variations between the C phase and the other two phases at high frequencies, appearing to be different from the others, which is also confirmed by the linear scale comparison in Fig. 11.3. This, at the first glance, can be assumed to be due to a specific winding design or, in some cases, poor grounding measurements.

Therefore, the same type and capacity sister unit with a successive serial number, having been assumed to be in a good condition, was also tested and its responses are taken into account to determine the overall diagnosis for the suspect unit. As seen in Fig. 11.4, the sister unit, assumed to be in a good condition, possesses similar variations at high frequencies between phase responses similar to the investigated transformer. This is an indication that the variations are likely to be related to specific design features [8].

In addition, a direct comparison of the two C phase responses of both the suspect and sister units shows good resemblance without unexpected deviations as shown in Figs. 9.5 and 11.5 using the log and linear frequency scales, respectively. The most distinguishable high frequency variations can be attributed to minor construction differences between the two transformers.

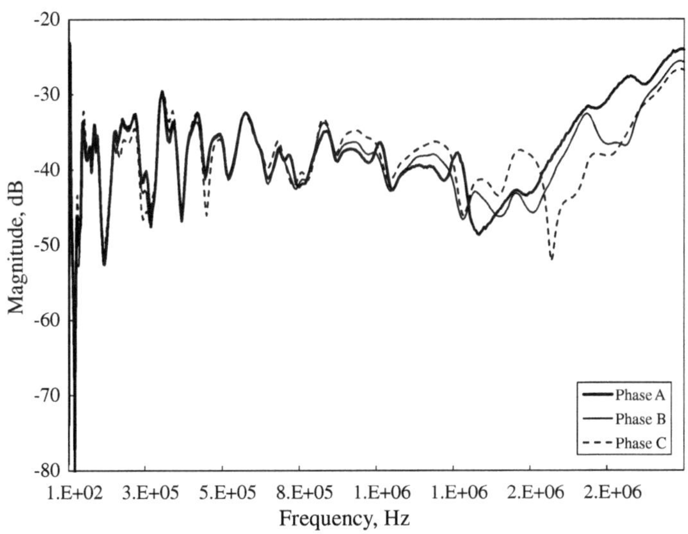

Fig. 11.3 "Construction-based" (phase) comparison (HV side, linear scale)

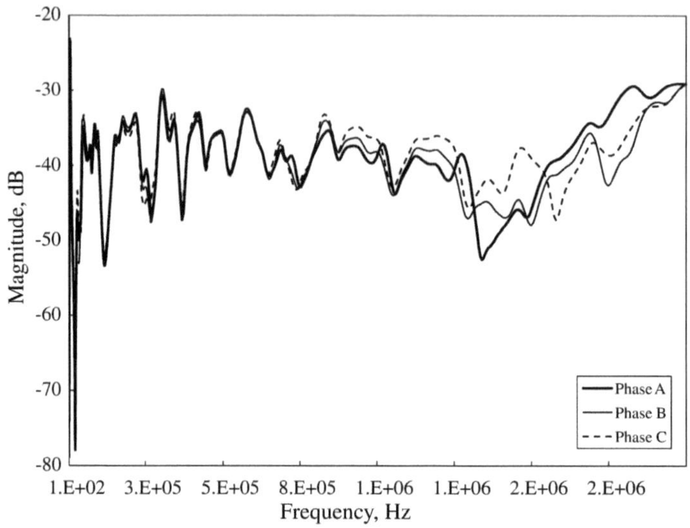

Fig. 11.4 Sister unit "construction-based" (phase) comparison (HV side, linear scale)

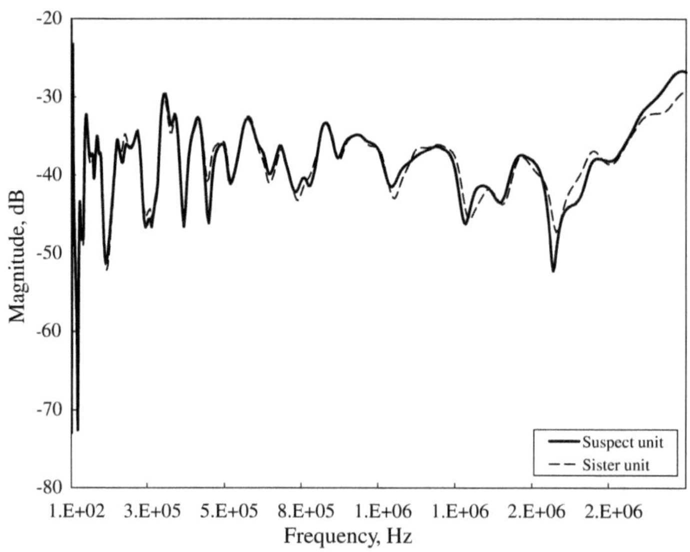

Fig. 11.5 "Type-based" (sister unit) comparison (HV side, phase C, linear scale)

Under the developed ER framework, the subjective judgements for the phase and sister unit comparisons can be represented, for instance, by the following distributions of degree of beliefs as defined in Eq. 3.29 using the evaluation grades as in Eq. 11.1:

Table 11.1 Subjective judgements for winding condition assessment of a 90/33 kV 75 MVA power transformer using the basic evaluation analysis model

Degree of belief ($\boldsymbol{\beta}$)		Weights (ω)	Hypotheses (evaluation grades)			
			NW (H_1)	SCT (H_2)	RM (H_3)	CF (H_4)
Basic attributes	RC (e_1)	0.6232	0.0	0.0	0.0	0.0
	PC (e_2)	0.1373	0.75	0.0	0.0	0.0
	SUC (e_3)	0.2395	0.95	0.0	0.0	0.0
Degree of belief ($\boldsymbol{\beta}$)		Weights (ω)	Hypotheses (evaluation grades)			
			HB (H_5)	AD (H_6)	PG (H_7)	
Basic attributes	RC (e_1)	0.6232	0.0	0.0	0.0	
	PC (e_2)	0.1373	0.0	0.0	0.15	
	SUC (e_3)	0.2395	0.0	0.0	0.0	

$$\begin{aligned} S(\mathrm{PC}) &= \{(\mathrm{NW}, 0.75), (\mathrm{PG}, 0.15)\}, \\ S(\mathrm{SUC}) &= \{(\mathrm{NW}, 0.95)\}, \end{aligned} \tag{11.3}$$

where only grades with nonzero degrees of belief are listed in the distributions. Note that subjective judgements in (11.3) are incomplete as discussed in Sect. 11.2. The assessment example is summarised in the decision matrix in Table 11.1, where the belief values for RC are assigned as zeros due to the unavailability of the reference (time-comparison) responses.

The objective of the evaluation problem is to obtain the overall diagnosis y using the available subjective judgements of e_1, e_2 and e_3, which constitute the bottom level of the basic evaluation analysis model for FRA assessment. Following the revised ER algorithm, presented in Sect. 3.2.2.2, basic probability masses are calculated using (3.30) and (3.31). Then by employing the recursive Eqs. 3.32–3.35 the combined probability masses for the three basic attributes are obtained. Finally, with aid of Eqs. 3.36 and 3.37 combined degrees of belief are derived and the overall assessment for the transformer winding condition is derived as the following distribution of degree of beliefs with regard to FRA diagnoses:

$$S(\text{Overall Condition}) = S(\mathrm{RC} \oplus \mathrm{PC} \oplus \mathrm{SUC}) = \{(\mathrm{NW}, 0.39), (\mathrm{PG}, 0.0216)\}. \tag{11.4}$$

In other words, there is a 39% probability that the transformer windings (and the C phase winding in particular) are in a normal condition and about 2.1% chance that there is *Poor Grounding* during the FRA test. However, from these results the remaining unassigned belief is $1 - (0.39 + 0.0216) = 0.5884$, i.e. a 58.84% of uncertainty in the final diagnosis. The relatively low probability of 39% for the *Normal Winding* condition and the large degree of uncertainty is explained by the fact that the responses for RC are not available and, therefore, the corresponding subjective judgements, which equals to zero, are included in the evaluation process as illustrated in Table 11.1.

However, if only the two basic attributes, PC and SUC, are considered without taking into account zero estimates for RC, the relative weight $\omega_1 = 0$ and, therefore, the first column and row of matrix (11.2) are eliminated. Thus, the corresponding relative weights are calculated using the reduced matrix (11.5)

$$\mathbf{A} = \begin{bmatrix} 1/1 & 1/2 \\ 2/1 & 1/1 \end{bmatrix}, \tag{11.5}$$

where $\omega_2 = 0.3333$ and $\omega_3 = 0.6667$. The final transformer assessment, using the reduced comparison matrix, gives the following distribution of degree of beliefs with regard to FRA diagnoses:

$$S(\text{Overall Condition}) = S(\text{PC} \oplus \text{SUC}) = \{(\text{NW}, 0.8960), (\text{PG}, 0.0267)\}, \tag{11.6}$$

illustrating an unassigned degree of belief of 0.0773 due to uncertainty in the expert's judgement. Thus, with about 39% and 89.6% probabilities of the *Normal Winding* condition in both the cases, it can be assumed that the transformer windings are most likely to be healthy.

11.4.2 Aggregation of Subjective Judgements Using the General Evaluation Analysis Model

In order to analyse a more general case when several experts are involved in an FRA assessment process, a complex diagnosis example is illustrated below. Suppose a transformer winding assessment task is carried out by three experts and their subjective judgements are listed in Table 11.2. The information is provided by the experts using the comparison techniques as discussed in the previous example and contains uncertainties. The assessment process can be represented using a general evaluation analysis model in Fig. 11.2 with three experts involved, i.e. $K = 3$.

Table 11.2 Generalised decision matrix for winding condition assessment using the general evaluation analysis model

General	Composite	Basic	Hypotheses (evaluation grades)						
Attribute levels			NW	SCT	RM	CF	HB	AD	PG
Overall evaluation y_1	Expert 1 (f_1) ω_1	RC (e_{11}) ω_{11}	0.15	0.0	0.1	0.3	0.4	0.0	0.0
		PC (e_{12}) ω_{12}	0.2	0.0	0.0	0.2	0.35	0.1	0.0
		SUC (e_{13}) ω_{13}	0.2	0.0	0.0	0.3	0.3	0.0	0.0
	Expert 2 (f_2) ω_2	RC (e_{21}) ω_{21}	0.3	0.0	0.0	0.25	0.3	0.1	0.0
		PC (e_{22}) ω_{22}	0.4	0.0	0.0	0.1	0.2	0.1	0.0
		SUC (e_{23}) ω_{23}	0.35	0.0	0.0	0.2	0.3	0.05	0.0
	Expert 3 (f_3) ω_3	RC (e_{31}) ω_{31}	0.2	0.0	0.1	0.3	0.4	0.0	0.0
		PC (e_{32}) ω_{32}	0.3	0.0	0.05	0.3	0.2	0.1	0.0
		SUC (e_{33}) ω_{33}	0.3	0.0	0.0	0.3	0.3	0.1	0.0

Table 11.3 Evaluation matrix for winding condition assessment using the general evaluation analysis model

Attributes		Hypotheses (evaluation grades)						
		NW	SCT	RM	CF	HB	AD	PG
Expert 1	(f_1)	0.1593	0.0	0.0711	0.2965	0.3954	0.0072	0.0
Expert 2	(f_2)	0.3298	0.0	0.0	0.2259	0.2962	0.0871	0.0
Expert 3	(f_3)	0.2239	0.0	0.0744	0.3047	0.3733	0.0204	0.0
Overall evaluation	y_1	0.2343	0.0	0.0438	0.2780	0.3711	0.0340	0.0

The relative importance of the combined experts' judgements (composite attributes f_k, $k = 1,2,3$) is assumed to be equal, which is represented by weight ω_k for the kth expert. Moreover, the weights of the basic attributes ω_{ki} ($i = 1,2,3$), being associated with the corresponding composite attribute f_k of the model, are assumed to be the same as in the previous example in Sect. 11.4.1. Thus,

$$\begin{aligned} \omega_1 = \omega_2 = \omega_3 = 0.3333, \quad \omega_{11} = \omega_{21} = \omega_{31} = 0.6232, \\ \omega_{12} = \omega_{22} = \omega_{32} = 0.1373, \ \omega_{13} = \omega_{23} = \omega_{33} = 0.2395. \end{aligned} \tag{11.7}$$

Since the assessment process is a multilevel hierarchical analysis problem, the first step of the ER analysis is to obtain the combined subjective judgements for each expert as discussed in the previous example. The calculated experts' combined assessments f_1, f_2 and f_3 are then aggregated to produce an overall evaluation y_1 as the overall transformer winding condition. The results are summarised in an evaluation matrix listed in Table 11.3.

As seen from the evaluation matrix, there is a large probability of 37.11% that the transformer winding is *Hoop Buckled* and a 27.8% chance of a *Clamping Failure* occurred. On the other hand, there is a 23.43% chance that there is no fault in the winding. In addition, there are small possibilities of *Residual Magnetisation* and *Axial Displacement* occurred. These results also imply a 3.87% of uncertainty in the diagnosis. As a result, it is more likely that the transformer experienced a winding deformation or a construction failure. Thus, it can be concluded that the winding of the investigated unit needs to be inspected.

The above two examples demonstrate that the revised ER algorithm used for FRA assessment is a simple and meaningful approach compared with the original ER algorithm in terms of computational complexity. The overall assessment of the studied transformer windings is presented as probability distributions of the occurrence of various failure conditions, which unifies and simplifies a decision-making process.

11.5 Summary

In this chapter an ER-based approach to transformer winding condition assessment is proposed to formalise winding evaluation processes using FRA. The proposed

ER framework is able to uniformly describe expert's subjective judgements with regard to various FRA comparison techniques. This aims to provide a balanced overall condition evaluation based upon collective expertise when one or more experts are involved. As the expert's subjective judgements are often incomplete and inconclusive, the revised ER algorithm is suitable to aggregate such original subjective judgements involving nonlinear relationships to determine the impact of uncertainty on an overall FRA assessment. The two examples have demonstrated the implementation process of the developed approach to ER-based condition assessment for transformer windings. Due to its open architecture, the ER evaluation analysis models can be flexibly adjusted to include new interpretation features and techniques. In conclusion, it is deduced that the proposed ER approach holds a potential for the formalisation of FRA condition assessment procedures concerning transformer windings, although much research is required to be carried out to obtain more reliable interpretation features with regard to different FRA diagnoses. This can lead to more reliable initial subjective judgements of experts.

References

1. Kim JW, Park B, Jeong SC, Kim SW, Park P (2005) Fault diagnosis of a power transformer using an improved frequency response analysis. IEEE Trans Power Deliv 20(1):169–178
2. Tang WH, Wu QH, Richardson ZJ (2004) An evidential reasoning approach to transformer condition monitoring. IEEE Trans Power Deliv 19(4):1696–1703
3. Yang JB, Sen P (1994) A general multi-level evaluation process for hybrid MADM with uncertainty. IEEE Trans Syst Man Cybern 24(10):1458–1473
4. Yang JB, Xu DL (2002) On the evidential reasoning algorithm for multiple attribute decision making under uncertainty. IEEE Trans Syst Man Cybernt A Syst Hum 32(3):289–304
5. Ryder SA (2003) Diagnosing transformer faults using frequency response analysis. IEEE Electrical Insulat Mag 19(2):16–22
6. Prout P, Lawrence M, McGrail T, Sweetser C (2004) Substation diagnostics with SFRA: transformers, line traps and synchronous compensators. In: Proceedings of the 2004 EPRI substation diagnostics conference, 22 pp
7. Ryder SA (2001) Frequency response analysis for diagnostic testing of power transformers. Electricity Today 13(6):14–19
8. McGrail T (2004) Use of sister units in SFRA analysis. Transformer SFRA application note 2003/10/06-01, Doble Engineering Co

Appendix:
A Testing to BS171 for Oil-immersed Power Transformers

Routine tests

All large oil-immersed power transformers are subjected to the following tests:

1. Voltage ratio and polarity.
2. Winding resistance.
3. Impedance voltage, short-circuit impedance and load loss.
4. Dielectric tests.
    a. Separate source AC voltage.
    b. Induced over-voltage.
    c. Lightning impulse tests.
5. No-load losses and current.
6. On-load tap changers, where appropriate.

Type tests

Type tests are tests made on a transformer which is representative of other transformers to demonstrate that they comply with specified requirements not covered by routine tests.

1. Temperature rise test.
2. Noise level test.

Special tests

Special tests are tests, other than routine or type tests, agreed between a manufacturer and a purchaser, for example:

1. Test with lightning impulse chopped on the tail.
2. Zero-sequence impedance on three-phase transformers.
3. Short-circuit test.
4. Harmonics on the no-load current.
5. Power taken by fan and oil-pump motors.

Index

B

Bayesian network, 51
- Bayes net, 50
- Bayes' theorem, 50
- Chain rule of probability theory, 51
- Conditional probability distributions, 51
- Conditional probability table, 52
- Directed arc graph, 155
- Likelihood, 50
- Posterior probability, 50
- Prior or marginal probability, 50
- Prior probability, 50

C

Computational intelligence, 9
- Artificial neural network, 108
- Bayesian network, 50
- Cybernetic techniques, 108
- Evidential reasoning, 38
- Evolutionary algorithms, 15, 17
- Expert system, 37
- Fuzzy logic, 48
- Genetic algorithm, 18
- Genetic programming, 24
- K nearest neighbour, 109
- Logical approach, 37
- Particle swarm optimisation, 29
- Support vector machine, 109

Condition monitoring and assessment, 2
- Dissolved gas analysis, 4
- Frequency response analysis, 5
- Partial discharge analysis, 6
- Thermal model, 4

D

Dissolved gas analysis, 4
- Bootstrap, 107
- Dörnenberg ratio method, 99
- Gas evolution, 96
- Gassing rate, 99
- Halstead's thermal equilibrium, 139
- Key gas method, 98
- Key gases, 95
- Relative partial pressures, 139
- Rogers ratio method, 99
- Total of all combustible gases, 95

E

Evidential reasoning, 38
- Analytic hierarchy process, 132
- Attribute level, 39
- Basic evaluation analysis model, 186
- Basic probability assignment, 40
- Basic probablity assignment matrix, 42
- Basic probability mass, 46
- Benefit attributes, 44
- Combined degrees of belief, 46
- Combined probability mass, 46
- Decision tree model, 127
- Cost attributes, 45
- Degree of belief, 46
- Dempster-Shafer theory, 40
- Evaluation analysis model, 39
- Evaluation grade level, 39
- Extended decision matrix, 39
- Factor level, 39
- Frame of discernment, 40
- Hypothesis, 40
- Local probability assignment, 43
- Multiple-attribute decision-making, 38

E. (*Cont.*)
Ordinary decision matrix, 45
Original evidential reasoning algorithm, 38
Overall probability assignment, 44
Partial combination rules, 42
Qualitative attributes, 39
Quantitative attributes, 39
Remaining probability mass, 46
Revised evidential reasoning algorithm, 45
Unassigned degree of belief, 47
Uncertainty, 37

F
Frequency response analysis, 5
Construction-based comparison, 170
Ladder network model, 178
Low voltage impulse method, 166
Sweep frequency response analysis, 167
Time-based comparison, 170
Transfer function, 165
Type-based comparison, 170
Fuzzy logic, 48
Crisp set, 48
Fuzzy set, 48
Membership function, 48

G
Genetic algorithm, 19
Crossover function, 21
Fitness function, 22
Initial population, 22
Mutation function, 21
Roulette wheel selection, 21
Selection function, 20
Simple genetic algorithm, 20
Termination criteria, 22
Genetic programming, 24
Crossover, 27
Fisher's discriminant ratio, 28
Fitness function, 28
Functions, 25
Mutation, 27
Population initialisation, 25
Reproduction, 27
Selection pressure, 28
Terminals, 25

M
Mathematical optimisation, 16
Constraints, 16
Decision variables, 16
Deterministic algorithms, 17
Enumerative schemes, 16
Objective function, 16
Random search algorithms, 17

P
Particle swarm optimisation, 29
Active aggregation, 31
Best previous position, 30
Global best position, 30
Particle, 30
Particle swarm optimiser with passive congregation, 31
Passive aggregation, 31
Passive congregation, 31
Position, 30
Social congregation, 31
Standard particle swarm optimiser, 29
Swarm, 30
Velocity, 30
Power transformer, 30
Copper loss, 1
Core loss, 66
Heat run test, 79
Stray loss, 66
Transformer fault, 2
Transformer thermal dynamics, 62
Winding temperature indicator, 67

T
Thermal model, 4
Comprehensive thermoelectric analogy thermal model, 61
Equivalent heat circuit, 63
Fourier theory, 61
Ohm's law, 61
Oil natural air natural and oil force air force cooling, 63
Simplified thermoelectric analogy thermal model, 68
Thermal capacitance, 66
Thermal conductance, 67
Thermoelectric analogy theory, 60
Thermoelectric analogy thermal model, 63
Transformer fault, 2
Arcing, 98
Corona, 98
Overheating of cellulose, 98
Overheating of oil, 98
Partial discharge, 6
Winding deformation fault, 164

Transformer thermal dynamics, 62
  Bottom-oil temperature, 56
  Forced oil cooling, 59
  Natural oil cooling, 59
  Oil directed cooling, 59
  Steady-state temperature, 56
  Thermal time constant, 57
  Top-oil temperature, 56
  Transient-state temperature, 56

W

Winding deformation fault, 164
  Axial displacement, 173
  Clamping failure, 172
  Hoop buckling, 173
  Normal winding, 171
  Poor cable grounding, 173
  Residual magnetisation, 173
  Short circuited turns, 172

MIX
Papier aus verantwortungsvollen Quellen
Paper from responsible sources
FSC® C105338

If you have any concerns about our products,
you can contact us on
ProductSafety@springernature.com

In case Publisher is established outside the EU,
the EU authorized representative is:
Springer Nature Customer Service Center GmbH
Europaplatz 3, 69115 Heidelberg, Germany

Printed by Libri Plureos GmbH
in Hamburg, Germany